THE DIGITAL SHIFT

CHRISTIAN VON REVENTLOW & PHILIPP THESEN

STEIDL

First edition published in 2019

Authors: Christian von Reventlow and Philipp Thesen
Contributing Editor: Thomas Huber
Design and Art Direction: Winkreative AG
Photography: Vincent Fournier
Illustration: Matt Blease
Portrait illustrations: Yuschav Arly

Scans and separations by Steidl image department
Production and printing: Steidl, Göttingen

Steidl
Düstere Str. 4 / 37073 Göttingen, Germany
Phone +49 551 49 60 60 / Fax +49 551 49 60 649
mail@steidl.de
steidl.de

ISBN 978-3-95829-580-3

Printed in Germany by Steidl

INTRODUCTION

The digital revolution is changing the world as we know it. Previously unimaginable technological possibilities are already in motion – or just around the corner.

Many decision-makers in industry and commerce are confronted with a fundamental question: which path should they choose for the development of their companies and brands? The engineers are ready. But are the citizens, consumers and customers ready too?

With this publication we provide an inspiring management tool for CEOs, business leaders and designers across multiple sectors, with a focus on tech- and design-driven companies. It will give readers practical examples and guidance. And it will showcase ideas regarding the implementation of fruitful innovation and design strategies that readers can apply within their own organisations, no matter what their industry.

This book covers:

1. The impact of Artificial Intelligence today and the need to humanise it
2. The essential role of the user experience in transforming Artificial Intelligence into what we will introduce as Personal Intelligence
3. How consumer insights act as drivers of transformation
4. The importance of scenario thinking and the need for "more fiction in science"
5. The changing role of design in product development and business strategy
6. The Personal Intelligence of tomorrow – and the infrastructure required to enable it
7. How to create a creative environment, and why it's vital to do so

Each chapter opens with a summary of "top actions" for the CEO, allowing them to quickly understand concepts and derive action plans. For readers interested in further details, we then present in-depth case studies. And we end each chapter with an essay of reflections and a more detailed action plan.

REFLECTIONS ON ARTIFICIAL INTELLIGENCE: IMPACTS ON BUSINESS AND SOCIETY

The central issue of this book is the rise of Artificial Intelligence. It will determine our future. The crucial question is: will it serve (and be of service to) humans, or will it rule almighty over us? To steer Artificial Intelligence in the right direction, we have to develop a compass that leads us along the right path in the digital development of our businesses, regardless of technology, industry sector, or the markets in which we operate.

With the rise of digitisation and Artificial Intelligence, technology permeates all areas of life. For decades, countless technological futures have been imagined and visualised. Literature and Hollywood films have given rise to romantic yet critical cyborg visions. However, as the human-machine continuum becomes more and more digitised, real life is set to truly change. How work and home life blend is the most relevant issue with regard to this digital transformation. When society begins to run out of jobs due to automation, the human race will face radical change.

The consequences of this digital transformation on the economy are evident. We stand on the threshold of the fourth industrial revolution. These are historic times. Driven by the internet, the real and virtual worlds are merging. Cyber-physical systems have appeared as part of a globally networked world where products, devices and objects interact with one another via the Internet of Things. Using sensors, these systems process data from the physical world and make it available to intelligent, network-based services. These, in turn, can directly influence processes in the physical world via actuators.

Some industries in Europe and all over the world are great at mastering this unique blend of the physical and the digital. Airbus, for example, has developed a digital twin for each part of its aircraft, such as wings and turbines. An entire physical system can now be represented in data, and anyone can use this data for their own application or simulation, networking and contextualising it with data in the cloud. The product memory, all spare parts, all interactions and measurement data along the entire lifecycle are stored in the digital twin. Thus, the aircraft does not only keep a journal about itself but it also receives a digital representation of its own distinctive "personality", distinguishing it from all other Airbus aircraft. The aircraft becomes an individual.

We are good at humanising technology by giving robots faces. We talk a lot about the Internet of Things and the Internet of Services. But what about the Internet of Users? Why do we have such good command over digital representations of machines, but have yet to develop a veritable digital human twin? A digital twin able to represent human beings in their physical entirety, containing the complete biography of their social, cultural, political and economic interactions? A digital twin that contains the complete history of their user experiences, in all imaginable constellations?

Silicon Valley tech companies have had digital twins of billions of citizens on hand for some time now. However, these digital twins are what could be referred to as "dark twins". Since these companies are exclusively interested in their consumer preferences, they only represent partial aspects of humans. Facebook and Google utilise Artificial Intelligence only to capitalise on gigantic data sets for advertising. We are dealing with pure media companies, networking, intelligently analysing and exploiting their users and their data. Within this logic, the customer is slowly becoming the product. Amazon is already one step ahead and has become more than just a gatekeeper to user-relevant information. Amazon has privatised the entire customer journey, from search and feedback forums to algorithm-based help in making purchasing decisions.

All these companies have one thing in common: like intelligence agencies they secretly collect and exploit data. If humans happen to encounter their own dark data twins, would they recognise themselves?

As a first step, it's a challenge to bring our dark twins out of the shadows. Citizens all around the world want to take back control over their data. There are clear majorities, especially in Europe, where people insist on tighter data security and stronger privacy protections. Those eager to avoid regulations must come up with satisfactory technological alternatives. Citizens are also consumers and customers. They must reclaim their data in order to regain consumer sovereignty. What they truly want is to decide for themselves which brands and companies they wish to interact with. We believe that companies which embrace consumer data sovereignty will prevail in the near future.

There is another reason for our growing unease about the rapid advancement of machine learning and Artificial Intelligence: the fact that we are not actively involved in the development of these fascinating technologies. Instead, we perceive ourselves as passive objects who are scanned and analysed to provide data fodder for improving AI's machine learning algorithms.

This is why dystopian fantasies created by Hollywood films and criticism against the uncontrolled application of Artificial Intelligence by voices such as the late Stephen Hawking and Elon Musk are so popular. We don't all share these pessimistic assessments and predictions of the future, but we must make considerable efforts to reconcile humans with machines if we want to increase societal acceptance for these technological achievements.

We will only succeed if we change our priorities. We do not wish to make machines appear harmless. Instead, we want to increase transparency regarding the effects of new technologies on our society. We should avoid reducing customers to products and instead celebrate them as sovereign consumers. We should not turn humans into passive objects of rapid economic and social changes driven by digitisation. On the contrary. We need to enable them to become self-determined agents in the useful application and smart use of these wonderful technologies.

This digital transformation of the economy, including all of its business models and products, could serve as the foundation for a new, consistently successful global prosperity paradigm. But in order to achieve this, the industry must finally start thinking radically from a human perspective.

We're talking about nothing less than the comprehensive humanisation of technology. Artificial Intelligence must become civilised. It must be brought to a level of social, cultural and moral development that chimes with current social progress. The relatively recent debate about machine ethics with regard to autonomous vehicles and autonomous weapons shows that we are only beginning to explore this field.

Many might object that the demand to domesticate the machine sounds quite archaic. However, one thing must be clear: the digital dog should never be allowed to bite its owner, or else the plug will be pulled.

In the long history of technology, all human inventions have been extensions of the human body and its skills. Numerous technological developments have reacted positively to human limitation: the hammer allows humans to extend force, automobiles allow us to travel further and faster.

We must not forget that, for it to remain valuable, technology must be created with people in mind. Consider the history of computers. When they were initially created, they existed as huge, electronic, mainframe devices, sometimes filling whole rooms. Then, as technology progressed, we were able to make them smaller and completely alter their appearance. They broke free from huge cabinets and, in the guise of personal computers, settled on our desks. The computer transformed from being a heavy device for large-scale industrial application into a light and ubiquitous tool for personal and private use.

Artificial Intelligence must complete a similar transition, to what we call Personal Intelligence. Whereas it is currently a tool used only by the elite, Artificial Intelligence must become a tool for everyone – a technology that helps, not hinders humans from becoming sovereign agents and users who can profit directly from technological progress.

Personal Intelligence stems from applying methods derived from user experience processes to Artificial Intelligence in a radical way. With User

Experience (UX) we have a large toolbox at our disposal. UX enables us to look at things from the perspective of the user in a new and helpful way.

We propose a new formula for Personal Intelligence (PI):

AI + UX = PI

In the near future, almost everyone will have a digital twin consisting of all their user experiences and interactions with digital services and products. The digital twin will represent the unique choices and tastes of every single user. Brands will speak to individuals directly, changing their offer depending on any one person's specific needs. When this happens, brands won't appeal to huge demographics, as they have done previously, but will target customer segmentations of one: the individual.

As a consequence, the discipline of design will take on a prominent role in the personalisation of Artificial Intelligence. Designers know people's desires, fears and actual needs. Designers shape the intuitive relationships of humans to the objects around them and to the digital representations of their interactions. That is why the discipline of design must take on a crucial societal role: it must humanise technology.

Innovators and designers will become mediators between technology and the living environment. As moderators within companies, they must mediate between the various entities involved. They must ensure that all development processes look beyond what is technically feasible: the consumer's needs must come first. As the consumer's advocate in our industry – an industry largely driven by engineering – the designer is in charge of empathy, of understanding what people need and creating ways to best meet their requirements.

Should the vision of a "digital twin" as a personal, digital representation of human interactions prevail, we will need a different kind of internet. A next generation, multidimensional internet must be designed from the ground up. Large parts of the two-dimensional internet must be built anew.

Personal Intelligence will make high demands on technological infrastructure. When humans integrate tools of Personal Intelligence into their bodies and thus autonomously extend their own skills, bandwidth and computer performance will have to be multiplied enormously and made available in close proximity to the user. Issues such as low latency and edge computing will determine the competitiveness of infrastructure companies such as Deutsche Telekom, since they represent the precondition for all of humanity to have equal access to the new tools. By then, connectivity will have become a human right.

Christian von Reventlow, PhD

Christian von Reventlow is focused on transforming organisations' culture, core processes and governance to create value through innovation. He wins Board support for the mission, operates outside-in starting with customer value and drives enterprise-wide deployment of the latest digital technologies such as AI, AR, IoT, digital twins and edge computing. Christian believes positive global impact can be created by applying technology with humans and the planet in mind. His career spans two and a half decades of management- and officer-level assignments at global technology companies including Deutsche Telekom, Intel and HERE. com (Nokia/Microsoft), creating multiple startups. His leadership profile is marked by a collaborative approach at the executive level and a dedication to managed personal accountability at the operating level.

Prof. Philipp Thesen

Philipp Thesen is a prominent pioneer in the field of strategic design for digital transformation. He helps organisations to implement design as a driving force, to distinguish the brand at the product and service level, and as a mindest for innovation and organisational transformation. Philipp has two decades of experience as designer, innovation consultant and leader of international design teams. As Chief of Design at Deutsche Telekom he was responsible for the design strategy, the design of all products and digital user experiences, and the worldwide implementation of customer experience design. His focus on design as a strategic asset for the business won hundreds of international design awards. Philipp is a professor of design with a focus on Human-System-Interactions at the Darmstadt University of Applied Sciences where he is also researching design and AI.

CHAPTER 1

IDENTIFYING THE DRIVERS OF DIGITAL TRANSFORMATION

The boundaries between the virtual and material worlds are disappearing – in industry as much as in everyday life. As computing power and the ability of Artificial Intelligence increases, formerly static and non-animated objects will come alive. The networking of virtual with material things will create novel and unexpected interactions, and lead to unforeseen behaviours and consequences. New patterns will emerge. New forms of consciousness will appear. People will develop needs we cannot even imagine today. As a result, people's relations to things and the environment will significantly change.

This new reality poses big challenges for business. CEOs and design and innovation leaders will have to identify new consumer needs and adapt their strategies accordingly, in order to develop suitable new products and digital services. To achieve this, it will be necessary to grasp novel technical possibilities, but the real key will be to understand the human aspect of this upcoming transformation.

We need to understand what drives emotionally engaging, intuitive relationships between people, their things, their environments and the digital representations of themselves. Throughout the book, beginning in this chapter, we will highlight a number of hands-on examples. These are reflections of the broader implications, and a practical implementation guideline.

Top actions for the CEO:

1. **Send people out into the world to share prototype experiences of future technologies, because mere graphs and statistical data on technological change have no persuasive power.**

2. **Uncover human aspects by engaging with customers in entirely new ways, because it is impossible to predict behaviour patterns through a linear extension of past experiences.**

3. **Create emotionally engaging discovery experiences for your company executives, because a purely rational approach will fail to grasp true innovation.**

CASE STUDY: THE KIDS ARE ALL RIGHT

One day towards the end of 2015, a small group of sociological researchers flew from New York to Helsinki and walked into a city centre school. It was their first stop on a world tour of classrooms. In the coming months they would fly to Amsterdam, Frankfurt, Seoul, Shanghai and Singapore. Later they would make four additional stops around Germany. And in each city, they would talk to groups of 10–17-year-olds about their digital habits. In what ways do they use technology? How has it affected their behaviour? What do they expect from the future? What needs do they have that are currently unmet? The researchers were there to understand what these children thought the world might be like in 2025, then a decade away, and what they might require in order to navigate it successfully. To put it another way: they were there to preview the future.

In the brightly coloured Helsinki classroom, the researchers quickly got down to work. First they asked the children to discuss topics they believed would be relevant in 2025 – privacy, connectivity, cool personal robots. Next, they asked them to consider what might change in those areas over the next ten years.

Prior to the workshops, the researchers had labelled the group "Digital Natives", a catch-all term for children born in this millennium, and quickly distinct patterns of belief specific to their generation began to emerge. The research team knew this to be the first peer group to grow up entirely networked, but they had underestimated to what extent living in a world of abundant information and fluid connections had affected their behaviour.

Out of the discussion came remarkable provocations. To the digital native, for example, privacy as the rest of the populace understands it does not exist. After all, their lives play out online. And yet they care about who has access to their vital statistics. They do not want their details available to everyone. And they will quickly swing towards a product or service if they know it will treat their data with care.

The team soon recognised other generation-specific behaviours. They noted, for instance, that they have a wildly intimate relationship with technology, to the extent that the traditional boundaries society draws between human and computer are meaningless to them. (A 2015 report by the Pew Research Center included the line: "79% of young people display symptoms of emotional distress when kept away from their personal electronic devices.") They discovered that anything that interferes with the digital native's ability to enjoy a frictionless experience – a server lag or a screen freeze for example – is considered an outrage. Above all, they came to understand that digital natives consider connectivity a human right.

Each workshop took place over half a day and all culminated in show and tell presentations. Partway through each session, the children were asked to create prototypes of societally positive products or services they believed would exist in the future. This was the fun bit: scraps of coloured paper, pieces of felt and tissue, pipe cleaners and plenty of glue. Out of standard classroom craft materials appeared crude facsimiles of future-facing products and services.

The children were asked to stand up among their peers and pitch their ideas, to share with the group what their solutions were and how they might affect life in 2025. The results were exhilarating, authentic, incisive and, in many cases, completely unexpected. We had conceived these workshops as a kind of experiment, but by all measures, the outcomes inspired awe.

*

We first discussed the idea to ask children questions about how they envisioned the future at the beginning of our journey. It was radical in two important ways.

To understand the first, you must consider the history of the telco industry. Since its inception, telco has evolved at astonishing speed. New products and services, facilitated by rapidly emerging technologies, launch weekly, often with disruptive effects. If they fail to offer what their customers need or demand, large organisations can begin to lose market share. There is a constant danger of falling behind.

What does the future look like through the eyes of children?
Two examples from the workshop's futuristic and colourful creations.

For that reason, big businesses tend to commission innovation studies that stick to providing short-term insight. They look at what their customers need now and six months in the future, meet those requirements effectively, and steer clear of making strategic calls on the long-term impact of new forms of innovation, lest they make a big mistake.

Before we arrived in our roles, Deutsche Telekom had attempted to break that mould. It had commissioned reports that began to examine trends and behaviours far into the future, but often the studies were targeted around existing products or lines of research. (One sample report asked the question: What will the future of the smart home look like in five years?) No broader, holistic angle was being taken. More worrying was the fact that the studies still relied on data collected from the customers of today, the same people who in ten years might well have shifted out of Telekom's target demographic. Talking to today's customers is great when you want to get information about today. But, as time passes, that information becomes less and less useful.

Before long, we raised an important internal question: To truly understand what Telekom needs to accomplish in order to remain a relevant player in the 2025 ecosystem, who do we need to speak to? The answer was simple. We needed to talk to members of society whom, in 2025, will be interacting with our products and services – and not just those who use only connectivity. Those people just happened to be kids.

In the IT and software industry, ten years can change everything. When developing strategy, very few organisations look that far ahead. That made our idea radical. But it was radical in another way, too: the focus of this study were children! Nobody in our industry had extensively reviewed the desires and future requirements of digital natives, despite the fact that they would soon become our primary customers.

For many within our organisation this concept was difficult to accept. When presented with the idea, individuals we spoke to within the company thought it was odd. They were sceptical of its potential outcomes – what if what the children said turned out not to be true? They are just children, after all – and worried about wasting money. Questions were asked, chief among them: What do kids know?

This line of query makes sense, but it is defeatist. We replied: If an organisation strives to place customer experience at its heart – and if, above all, it wants to understand what it can do to remain relevant in the future – why wouldn't it ask its customers for their opinion, no matter their age?

Soon, we began to consider the kids' study as the potential driver of a new strategy for innovation. We proposed that it would form the basis of our future work, that it would show us the way. If this study could help us attain a firm perspective on the desires, requirements, interests and disinterests of our future customers, surely we could create a ten-year strategy that went some distance towards the creation of products and services they actually wanted, thus securing Telekom's position as an industry leader? This kind of insight, our thinking went, could drive transformation.

*

While travelling around the world, shifting from classroom to classroom, the research team began to draw a full picture of the future the digital natives envisioned. As the participants presented their prototypes, the researchers constructed a list of statements that aggregated the participants' beliefs about what will be meaningful in the future. They are as follows:

1. Everything will be networked.
2. Environments will become more intelligent.
3. Cities will respond to my individual needs.
4. Personal robots will make our lives easier.
5. Digital communication will blend into the physical.
6. Virtual and augmented reality will be the mainstream.
7. Actions will replace the interface.
8. Biometrics will be the new password.
9. Information collected about me, I control.
10. Technology will be an active enabler for new behaviours.

For each statement, numerous related prototypes were created. Sometimes those models resembled products already on the market, or at least they echoed concepts that existed in society's collective consciousness, perhaps through film or literature. A participant in Seoul, for example, created a pair of smart glasses (using pipe cleaners, tape and transparency film), onto which he'd drawn the schematic of a home, as though the user were assessing and interacting with its vital functions – the temperature, say, or its security system – from afar. Children in Amsterdam, Singapore and Frankfurt created similar prototypes. Of course, smart glasses already exist. Some of us even wear them. But for each prototype that resembled a product already available, plenty of zanier ideas also emerged. More than half of the children who took part in the study believed they'd be able to use similar glasses in entirely new and original ways. For instance, to experience a live concert, in 3D – *through the eyes of the pop star on stage*.

The zany ideas kept coming. In the future, public transport services will know who we are and where we want to go, no matter where that might be.

Home environments will recognise our mood and react accordingly, perhaps by dimming the lights. We will own or employ personal robots, some of which will be able to decipher emotion and desire. We will talk to people in 3D – holograms were the most referenced form of future communication in the entire study – enabling intimate but remote conversation and emptying our roads of cars, which may soon be considered impractical. Our beds will scan our bodies and measure our vitals, sometimes urging us to visit the doctor or change our diet. Rings we wear on our fingers will act as mobile wi-fi networks, as will belts. Our sunglasses will warn us of prospective environmental dangers. We will control the amenities within our homes with gestures. We will be able to project digital worlds onto physical spaces. We will access bank accounts with our retinas. And our personal data will be stored in a single place, its control remaining firmly in our hands.

Ultimately, technology will exist to enable us, in ways few of us could ever have imagined.

As each prototype was shared, the research team added information to what quickly became a great wealth of data, all of which revealed the participants' desires, fears and hopes. Insight! They now knew what digital natives considered the most pressing issues society would face in 2025 – digital privacy – and what mattered far less. They found out what people might wear, how we might communicate, how we might keep track of appointments, or keep ourselves safe. And they began to understand the importance of connectivity among digital natives. All of it pointed to a new form of human-centric technology – one that had our personal interests at its core. The only thing left to do was present the findings back to Telekom.

*

The transformation of a large organisation is not an easy task. Deutsche Telekom employs more than 200,000 people around the world. There are rules and regulations, hierarchies and chains of command. Innovation is a Telekom bedrock, but it can be usurped by other interests. But once it had been assembled and shared, the kids' study became a major internal success. Key Telekom stakeholders approached our team and invited us to share our findings with larger groups. Word of the report spread and momentum gathered. The whole idea was thought of as new and novel; few employees had considered – or even been allowed to consider – the requirements of customers so far in the future. In some cases, our findings directly impacted products and services launching to market, and allowed us to develop the foundations for future projects. The insight we gathered actually began to drive transformation within the organisation.

In many cases, the report supported paths Telekom was already pursuing. Take Augmented Reality. Telekom products incorporating AR were already being developed, but our findings validated their relevancy, and reinforced the requirement for further investment. But in some instances the study has initiated all new realms of innovation. The kids were crazy about robots, for example, so Telekom is now experimenting with robots. And the children very firmly believed connectivity to be a human right, so Telekom is beginning to face the prospect of a future in which its primary product is available for free, or at least available in a kind of tiered system – connectivity might soon be available gratis, while premium connectivity might come at a cost. And all of this in a post-mobile world, no less – a scary prospect, given Deutsche Telekom has about 150 million customers, but one worth being aware of sooner rather than later.

But the study had another purpose: it helped us clarify our strategy for the next decade. It provided us with a set of priorities and allowed us to focus on the areas of innovation we needed to explore. Now we had created a firm and common understanding of the direction we needed to take: we would pursue areas of innovation that related specifically to what the children had spoken about. We named the model 4+1, which referred to four short-term projects that created innovation value within Deutsche Telekom and involved connectivity, the Internet of Things, the smart home and Artificial Intelligence-focused customer service.

Imagine, for a moment, the ramifications of the study. Here is one of the world's largest telcos adapting its strategy for innovation around the thoughts of a group of international 10–17-year-olds. And yet the approach has been incredibly successful. The study has yet to infiltrate all areas of the organisation. Perhaps it never will. But the report has at least enabled a large degree of internal preparation. It has, in a way, future-proofed the organisation. Nobody can claim that Telekom will be surprised about the kind of world we all will be living in in 2025. The kids slipped us the necessary intel long ago.

ESSAY:
KNOW THYSELF (AND THY FUTURE)

Technologists have long understood that the rate of technological change is not going to slow. In fact, the opposite is happening. Technology is advancing at an exponential pace. The changes we have seen in the past ten years have been monumental, but they will be comparably tiny to those that will be made in the next ten years and they will be nothing in comparison to the decade following that one. As technologists such as Ray Kurzweil have argued, we have entered the second half of the chessboard of exponential technological growth. The world will continue to change at an astronomical rate. And so will we. It's not an exaggeration to suggest that the people we are today would probably fail to recognise the people we will soon become.

Specifically, the rise of Artificial Intelligence will determine the future of humanity. But which direction will this future take? Is the increasing digitisation of our economy and society actually improving our lives as the old mantra of progress has been suggesting for a long time? Or will all present certainties evaporate under the pressure of ever-accelerating change? One thing is clear: the digital transformation does not only concern the economy. It profoundly affects people's lives and their very identities.

Digitalisation concerns us all. It transforms the economy, society and democracy. It will determine how we will live and work in the future. It will mean that people will develop completely new needs. Needs we do not yet know. To retain relevancy, innovation and design leaders should begin to question exactly how our everyday lives will be transformed. And they should pursue new strategies that adapt to this reality and to develop suitable new products and digital services.

As these two worlds merge, embedded electronics and software systems will play an important part as relevant drivers of innovation. They will significantly expand the functionality and therefore the practical and added value of vehicles, aeroplanes, medical devices, production facilities and appliances. Cyber-physical systems will emerge as part of a future in which products, devices and objects are connected across the globe and interact with each other. As the man-machine interfaces become more diverse, humans will merge with the virtual world of data, things and services. And it will have profound consequences on the notion of individuality – what it means to be a human being.

But today's technological developments are only the beginning. The expected quantum leaps in the development of Artificial Intelligence, which we will experience within the five years, will raise even larger questions. What will happen if machines become more human, and the human becomes a tool used by machines?

In the course of these shifts in the man-machine continuum, our lives will change. While debates about the digital transformation often solely focus on aspects of industrial policy, the relevant topic for us is the fusion of work and life outside of work. The consequences of such a fusion aren't predictable. It remains to be seen if this path will lead to freedom or, on the contrary, to a complete loss of control over one's own life. The debate about the social and political implications of the digital transformation is in full swing. How can we balance human needs and technological possibilities?

To address these thorny issues and find fresh solutions, it is necessary to apply methods of scenario development and design thinking. They can help create new products and business models from the intersection of the human and the technological, generating new opportunities on the market and improving customer loyalty. To do so, innovators need a mindset that is solution-oriented rather than problem-oriented. Creative fantasy and imaginative power are needed to gauge the future and to construct a development trajectory that corresponds with the changing needs of customers and consumers.

These considerations raise a series of important questions. In what ways will the digital transformation alter who we are and the way we live? How will it change the jobs we do and the way we do them? How will it affect the way we communicate? Or travel? Or eat and drink? How will it alter the skills we deem important or the languages we speak? Those who understand how technology will affect the future will better serve our needs in meaningful ways. Those who don't will fall behind.

This is particularly pertinent in business. Let's identify why. For the past 50 years, organisations around the world have created long-term business strategies based on the events of the past. For decades it was considered safe to assume that what happened last year will happen again this year, and the next. That made strategy easy. Imagine the thought process of a CEO. She thinks: "The business strategy we applied last year was successful. We approached the market in the correct way and achieved our targets. In the meantime, the world hasn't changed much, so we'll use the same strategy this year, too." Often, that same approach was put into practice again and again, year after year, only requiring a tweak now and then. What's the point of changing a successful process, if the same thing happens every year?

But the past can no longer be relied upon as an accurate predictor of the future. Today, we live in a world characterised by fundamental volatility, uncertainty, complexity and ambiguity. The exponential growth of technology has accelerated the rate of change to such a degree that even the near-future is almost impossible to predict accurately based on past events. What happened in 2017 will be vastly different to what will happen in 2020, for example. For all we know, 2030 will be unrecognisable.

HOW TO IDENTIFY DRIVERS OF TRANSFORMATION

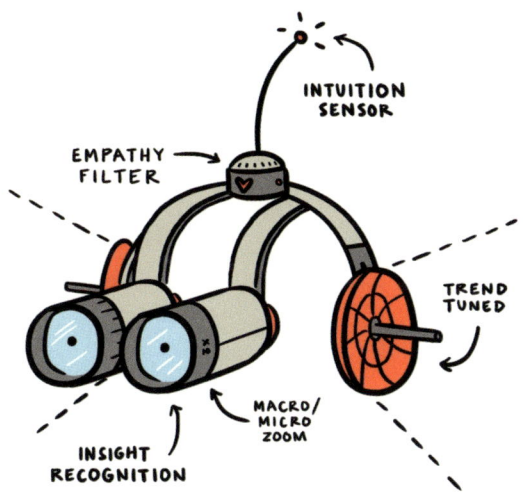

INTUITION SENSOR

EMPATHY FILTER

TREND TUNED

MACRO/ MICRO ZOOM

INSIGHT RECOGNITION

DRIVERS OF TRANSFORMATION IDENTIFYING GOGGLES

Identifying drivers of transformation requires empathy. To grasp change, you need intuition, an awareness of trends and cultural sensitivity, as well as a profound understanding of the true wishes and needs of target groups. A deep knowledge of the lifestyles, living conditions and cultural discourses that are significant among identified target groups is especially valuable for the innovator. Without the skill to listen, it would be impossible to create relevant practical value and meaning. Innovation managers and designers can here be of significant help to articulate the needs of customers and integrate them into solutions. Therefore they can go beyond the technologically feasible and give a product essential meaning, taking into account the intuitive relations between people, their things, their environments and their digital representations.

Business schools will tell you this is easy: businesses have to understand who their consumers are, what they think and when, how they act and when, and what their needs are, both now and in the future. And they have to understand what is happening in the world, in order to achieve and maintain success. And this, they will tell you, can be easily achieved through focus groups, sociological studies and quantitative research.

Will this approach work in the future? No. The exponential dynamics of the digital transformation expose the limitations of the classical tools, because they rely on the assumption of a linear progression of reality. The fluid environments of tomorrow, however, require new methods. To grasp these new realities, people will not be persuaded by logical arguments alone. The changes are so profound that past experiences of employees and executives become irrelevant. Instead, emotionally engaging experiences are required to create new mental models for future realities. As small pockets of resistance in organisations can block overall progress, this point becomes even more relevant.

Actions:

1. **Identify and engage accepted opinion leaders on all levels of the organisation. Don't be surprised if their number turns out to be up to 10% of the organisation.**

2. **Deploy classical research tools: combine macro, micro, trend, technology research and derive a variety of scenarios. Both qualitative and quantitative.**

3. **Send people out into the world to share and discover prototype experiences of future technologies. Because mere graphs and statistical data on technological change have no persuasive power.**

4. **Uncover human aspects by engaging with customers in entirely new ways. Because it is impossible to predict behaviour patterns through a linear extension of past experiences.**

5. **Fuse the results through a multitude of events involving the opinion leaders above. Distil the top five drivers.**

6. **Create emotionally engaging discovery experiences of the results for the executives. Because a purely rational approach will fail to grasp true innovation.**

Let's look at the recruiting industry as an example of what action a CEO might take in the future. Consider the head of a recruitment firm. He is aware that, within the next decade, Artificial Intelligence will completely transform the job market. AI will take over a large variety of tasks and jobs currently performed by humans. At the same time, a number of entirely new job roles will be created for humans. The recruiter understands that his industry is going to be unrecognisable within a matter of years, and yet he doesn't know exactly how it will change. So how is he supposed to react?

One answer is to commission studies. For example, applying the method from our case study, he could create a report collating the thoughts of individuals who will be entering the jobs market in ten years' time. Let's say those people will be aged between 6 and 14. They are a demographic that few people in the recruitment industry have ever bothered to question, because it is not a group that is working right now. But they soon will be. The recruiter can ask them questions that elicit thoughts about work, as well as their opinions on the future of technology. This will allow the recruiter to better understand the kind of jobs this generation is likely to create for itself, and he will know which technology he should use to target them when they begin to seek employment.

At the same time, the recruiter can commission a wider sociological study. This report would focus on the societal perception of Artificial Intelligence, and how we believe Artificial Intelligence will affect work. It will be valuable for various reasons. It might point to which jobs are likely to be taken over by algorithms, and which are likely to remain human pursuits. Or perhaps it will gauge how much of our lives humanity is willing to allow Artificial Intelligence to control. Think of the potential outcomes for the recruiter. Here's an example. What if the study points to the notion that governments around the world will soon limit Artificial Intelligence's influence over society? Won't the jobs market then be much less likely to change? Doesn't that help the recruiter better prepare for what's about to happen?

Insight won't just be required by recruiters. It will be important for any and all businesses affected in some way by technology. That means every business.

CHAPTER 2

DISCOVERING PATHS TO POTENTIAL FUTURES

Predicting the future is the foundation for strategy work. Exponential acceleration of technology makes predictions more difficult. Therefore, CEOs, business unit leaders and innovation managers need to create a set of potential futures as input for their planning work.

The biggest challenge is to identify unknown needs, ones that will emerge in the future and ones that are still latent. Users cannot yet express them, so the innovation manager must identify them. To do that, he needs to apply multiple strategies in parallel, hoping that one of them allows him to uncover the combination of an unmet need with its yet unknown technical solution. This approach is nothing new, but it has to be applied diligently.

The situation is further complicated by the adoption curve of technology. Many of the technologies that determine our lives today have been admired by previous generations only in science fiction films. What was once imagined by science fiction writers, from swiping on screens to gesture control and precrime (preventing murder by precognition), is now becoming mainstream.

However, the early adoption of new products and services cannot predict their mainstream success. New technologies might be deployed in many ways. These scenarios need to be discovered. In this chapter we will provide methods and examples of how to achieve this.

Top actions for the CEO:

1. **Kick off multiple activities to forecast the future in your industry and customer segments plus all adjacencies.**

2. **Spur creativity by playful discovery of seemingly unrelated items, because often the greatest inventions are just transfers from other disciplines.**

3. **Develop a sensory system to assess upfront how much change the organisation would embrace. Act wisely.**

CASE STUDY: YOUR MANY FUTURES

As part of an exercise at Deutsche Telekom designed to ascertain how the organisation might future-proof itself, we created a series of event cards on which possible future scenarios that might affect the telco industry were written. (The exercise was based in part on game theory.) The scenarios represented probable and improbable trigger events that might require a change in Deutsche Telekom's strategy. Here are fictional headlines that relate to two of these scenarios:

GOOGLE ACQUIRES SPRINT
Countering the acquisition of Microsoft by Verizon for $180b in the US in May, Google acquires Sprint for an undisclosed sum. Telco companies threatened by emergence of new VoIP provider.

And:

APPLE DISCONTINUES IPHONE
According to data published by research firm IDC, the global shipments of wearable devices has surpassed that of smartphones. This presents a radical shift in consumer habits, where customers are increasingly using wearables to communicate while smartphones are primarily used to view long-form content.

During the exercise, top Telekom executives were asked to discuss each event in small groups and decipher how it might impact the telco industry. Based on this understanding, each group sketched a set of potential futures for the telco industry, and came up with ways Deutsche Telekom might respond.

The findings were surprising. All groups landed on similar scenarios and offered similar responses. Soon we discovered that, together, we'd create a set of three potential futures. One was a linear extension of the today – it depicted a future as though none of the major events shown on the scenario cards had happened. The two others depicted futures that had undergone radical change following major disruption in the telco business, as though the two events described above had actually happened.

The top executives had produced impressive visions of the future of Deutsche Telekom.

But was this scenario analysis helpful? Ultimately, no, because the workshop, though exciting and fruitful, did not lead to any action. In retrospect, we believe that we failed to create the emotional engagement required to overcome the natural resistance to change that every organisation faces. We should have translated the scenarios into tangible practical experience. Plus, we failed to reframe the results so that they would fit the pre-existing mental frameworks of the participants. While we may all talk rationally, the reality of decision making is emotional.

Another approach in using scenarios was more successful. We invited a group of science fiction authors to Deutsche Telekom's headquarters to read excerpts from their work. The writers sat on a stage in front of a live audience and presented eccentric, often outlandish, texts about the future. Once the readings had ended, the audience was encouraged to ask questions about the concepts expressed, which sparked a number of interesting conversations, mostly dealing with the future paths of society – how will we all travel, work, communicate?

Due to its popularity, what was imagined as a one-off event quickly evolved into a monthly series. Soon, more and more sci-fi authors began to show up at Telekom HQ, bringing with them fantastic, prophetic visions. Members of the audience – Telekom designers, product managers, engineers and senior executives – listened with great care. Some of what was being talked about was similar to technology they were already developing, but often the concepts the writers had conceived were startling. In any case, a commonality appeared: the visiting writers shared a tendency with our team to think in scenarios so that they could imagine how technologies might soon affect our everyday lives.

One day, the event included a panel that featured the owner of the Phantastische Bibliothek Wetzlar, a library that holds one of the largest collections of sci-fi literature in the world. The library, founded in 1987, extends across five floors, holds 270,000 books and countless rare titles. Not just paperbacks, but periodicals, dissertations, magazines, underground zines and newspaper clippings. Visitors are invited to loan titles free of charge. You can find books by Arthur C. Clarke and Isaac Asimov, Philip K. Dick and Ursula K. Le Guin. Margaret Atwood's novels are there, as is work by J. G. Ballard and Octavia Butler, Kurt Vonnegut and Jules Verne. For anyone interested in speculative fiction, the library is a treasure trove.

At the event, the panellists discussed concepts plucked from decades-old literature that had become reality: personal computers, autonomous travel, gesture control. But they also discussed technological innovations that had yet to surface in the real world. Before long, it became clear that the audience was being provided with a series of scenarios that, when considered together, created an elaborate but theoretically realisable picture of the future.

Sci-fi authors are excellent storytellers. Like all writers, their great skill is to hook readers with fascinating narratives. But they are also excellent researchers, and many of the concepts they develop are based on science that exists. They embellish scientific ideas to a degree that often the objects and situations that populate their fictional worlds appear unrecognisable. But it is also true that it is not beyond the realm of reason to believe that the ideas they present would be possible to produce. In other words, almost everything a sci-fi author writes might one day be real, if only designers and engineers could make it so.

The discussion led us to an idea: what if we could compile a book that introduced some of science fiction's most interesting concepts in an effort to create discussion – and, eventually, prototypes – around future technological trends? Would the book's readers consider the ideas preposterous or insightful? Would it broaden their minds and inspire them to create, or would it lead them to dismiss the concepts as fantasy and speculation? And, most importantly of all: could science fiction be used to enhance our innovation strategy? Would it help our team envision the future?

We learned that science fiction writers have a special tendency to predict the future. In this way, many people believe that it is science fiction writers, not programmers, designers and engineers, who lead the world in innovation. True, it is science fiction writers who dream up new ways for us to eat, talk, get around and work. Their ideas shape the jobs we have and the kinds of fuel we use, the objects we buy and how they're made. Asimov and Clarke do not actually make the products and services we use, of course. (That's our job.) But the disruptive ideas that change and affect our world – autonomous cars, for instance – often come first in literature and only later in real life. "It's nice that science fiction exists," Palmer Luckey, co-founder of Oculus VR, once said. "Because these are really creative people figuring out what the ultimate use of any technology might be. They come up with a lot of incredible ideas."

*

As soon as we began the research for the book, our team found countless sci-fi examples relevant to the study, and we began to record ideas within five subject areas: Life, Living, Mobility, Communication, and Robots & Work. Important themes kept cropping up, many of which directly related to how we live our lives today.

For example, in a passage by the German author Wim Vandemaan, we discovered references to transhumanism – the emerging reality that technology can work in tandem with the human body to spur intellectual and physiological advances. (We read other authors on this subject, including Asimov, Clarke, Robert A. Heinlein and C. S. Lewis.) In the work of Christian von Aster, we found references to cleaning drones. In a passage by Frederik Pohl we read what appeared to be an early imagining of the Artificial Intelligence integrated "smart" bed. And in a book by Johann Seidl we found a passage about a "smart" fridge that refused its owner alcohol on account of data the fridge was processing about the state of the owner's kidney. That beer? Bad idea.

Later we discovered ideas relating to mobility, and also to robots. In the 2002 Richard Morgan novel *Altered Carbon*, we came across references to a transit clone that travelled great distances to appear halfway around the world in physical form, so its owner did not have to. In the Matthias Oden book *Junktown*, we stumbled across a passage that prophesied lanes and lanes of empty roads. Oden predicts that there will be little driving in the future, because there will be better ways to travel. In the Neal Stephenson book *The Diamond Age*, published in 1995, we came across a page where a 3D printer creates an entire personal spaceship in just a few hours. In the Ernest Cline book *Ready Player One*, he noted down a passage about a domestic robot that was able to simulate more than 2,000 smells.

We kept on reading and before long, we had a great list of innovations. There were drones and clones, instances of telepathy and self-driving cars, personal health scanners and those smart fridges that have their owner's best interests at heart. Some of the concepts we discovered were already out in the world, but many of the innovations had yet to be realised. We began compiling the information into a book. Each of the book's chapters centred around an essay that contextualised our findings. When pertinent, an entire passage from one of the library's books was included in full. Before long, a complete study had emerged. We called it *Science fiction als Inspirationsquelle für Innovation*, or: Science Fiction as Inspiration for Innovation.

The study shows the potential to impact innovation at Deutsche Telekom in much the same way the kids' study did: by utilising insight to affect real change. We see the concepts in the sci-fi study inspiring our own team to develop new products and services, many of which may well appear faintly familiar, at least to sci-fi fans.

From designers and engineers to product managers and senior execs, the book also serves as a role model for our employees. In many ways the work of an innovator is similar to that of a sci-fi author. They both share a common stock-in-trade: to imagine scenarios. But while the writer imagines scenarios and leaves it there, it is our role to select and bring such scenarios to fruition. Not only must we envision multiple new worlds but we also need to choose the ones we actually want to create. We also have to manufacture emotional engagement for those worlds we want to play a part in seeing come true.

ESSAY:
PUTTING MORE FICTION IN SCIENCE

STANLEY KUBRICK

T he biggest challenge for the professional working in the age of digitisation will be to identify the unknown and latent needs of their customers. To counteract this problem, it's very important to begin thinking in terms of scenarios. How will a new technology change our everyday life? How will it affect the way we live and work? Here, a lot can be learned for innovation managers and designers from the realm of science fiction.

Here's one example. When did the great Apple iPad first see the light of day? In April 2010? This answer is right and wrong. Such tablets could already be found on a console in the starship canteen in the classic 1968 film *2001 A Space Odyssey*. In a scene from this film, two astronauts, Frank Poole and Dave Bowman, watch a TV show on their tablets during their lunch break and the supercomputer HAL 9000 relays the fact it is content with being an artificially intelligent and autonomous specimen. This scene was screened by Samsung as part of a lawsuit in 2011, to highlight that it wasn't Apple, but the film's director, Stanley Kubrick, who invented the computing tablet. For designers today, this scene merely shows that the tablet came to the market nine years later than the year 2001 that Kubrick had imagined this innovation to exist.

How can it be that a technological innovation of today was foreseen almost 50 years ago? Is this just a one-off incident? Or is there something systematic about it? Was it just a case of brilliant science fiction authors accurately predicting the future? Or did their visions actually serve as blueprints for the engineers who simply translated literary fantasies into new technologies?

The famous chicken-egg discussion is one that science fiction authors have truly appropriated for themselves. If you were to evaluate all literature, you would see that writers have won this debate – purely from a statistical perspective. Many of the technological innovations that facilitate and determine our everyday life today were described in books written 30 or 40 years ago.

One particularly impressive example is the author Philip K. Dick. Up until his early death in 1982, Dick wrote about 120 short stories and more than 40 novels. He is considered one of the greatest science fiction authors of all time. Dick's stories are still relevant to the present-day reader. They make us question reality and the existence of humans and machines. How they would coexist, what there differences and similarities are.

While Dick's stories and theories were overlooked during his own lifetime, more than 30 years after his death the author is regarded as one of the most perceptive visionaries of the digital age. Dick was claustrophobic, he hardly ever left his house in middle-class Californian suburbia. Dick never did any research – the internet didn't exist back then – so he invented everything himself. In Dick's imagination, he travelled through countless worlds aided by extensive drug experiments.

Films like *The Matrix* and *eXistenZ* are based on Dick's ideas. This writer almost single-handedly defined the technological visions of the future that are portrayed by Hollywood. The list of film adaptations of Dick's novels and short stories is as long as it is impressive: *Blade Runner, Total Recall, The Man in the High Castle, Screamers, Paycheck* and *Minority Report.*

It has already been more than 15 years since Steven Spielberg created the film *Minority Report,* based on the short story by Dick. This film, with Tom Cruise in the lead role, captivated audience members with its numerous innovative technologies that were novel to the audience at the time: ubiquitous large-format digital displays, iris and facial recognition, and much more. Today, we can observe many of these concepts being close to come to life. For example, in experimental settings large-scale displays are able to recognise consumers standing in front of them. And displaying advertisements targeted specifically for whoever is standing there.

What we once admired only in films now determines many aspects of our real lives. The idea of operating a computer by mere hand movements, like in the film *Minority Report,* evidently inspired developers at Apple and Microsoft. Precrime, where police departments of the future use precognition to prevent murder, also used in *Minority Report,* is increasingly being utilised around the world. Today, numerous "Predictive Policing" models analyse case data with the help of Artificial Intelligence, calculating the probability of future crimes.

For a designer, the most impressive thing about *Minority Report* is the background story of its production development. At the start of the project, Steven Spielberg invited well-known futurologists and MIT scientists to a three-day seminar with the aim of finding out what the world might look like in the year 2054. This approach differs very little from the methods of design thinking, which designers use when describing the future. From the results of the seminar and the opinions of experts, Spielberg created a list outlining how medicine, transport technology, urban architecture, design, etc. are expected to develop by 2054.

When Spielberg was interviewed after the release of the film, he was quoted as saying, "I wanted to give the film some roots – get more science than fiction from it." This describes what we propose CEOs, business unit leaders and innovation managers do: Read science fiction and involve scientists to predict what the future might look like.

HOW TO DISCOVER PATHS TO POTENTIAL FUTURES

Actions:

1. **Kick off multiple activities to forecast the future in your industry and customer segments plus all adjacencies. Multiple activities are required, as it is unpredictable which activity will and will not create traction in your organisation.**

2. **Spur creativity by playful discovery of seemingly unrelated items. Because often the greatest inventions are just transfers from other disciplines.**

3. **Build scenarios using event cards of likely and unlikely business events to derive potential strategic actions.**

4. **Use design expertise to make scenarios accessible for laymen. And make them as tangible as a real experience.**

5. **Synthesise with discipline and care while paying attention to details. Structured thinking will help to separate the probable from the impossible.**

6. **Develop a sensory system to assess upfront how much change the organisation would embrace. Act wisely.**

7. **Create an experience lab like the Design Gallery while turning your clients into long-term creative partners.**

We'd like to reiterate something: to make scenario analysis work is extremely challenging. Imagining multiple and plausible versions of the future requires consideration of hundreds, if not thousands of variables. Often, it means challenging the standard assumptions and way of working of your industry.

Encourage your team to imagine scenarios no matter how crazy they might seem. Flying cars? It could happen. Artificial Intelligence integrated clothing? It has potential. Telepathic fridges? Why not! But how?

And how to take the organisation along?

By creating an experience lab that allows clients to expose themselves first hand to the innovations your organisation is developing, you are both showcasing your expertise and encouraging clients to become a co-producer of customised technologies and innovations. When clients gain a sense of ownership of an idea, they are much more invested emotionally as well as materially and the result of such a collaboration is usually very successful.

Whether or not a space like an experience lab exists at your organisation is up to the CEO. It is they who hold sway over such matters. If you are the CEO, we advise you to support a space like this to ensure your organisation is creating products and services of real value. If you are a senior executive, we suggest drawing up plans for a design gallery of your own, including a breakdown of how it will work, and how it will benefit your organisation. Then take those plans to the CEO and show them how valuable an asset this would be. The success of Telekom's Design Gallery is a case in point.

Let's look at an example of how futures can be made tangible:

The Telekom Design Gallery makes room for the future by bringing tomorrow's customer experiences to life. What does this look like in practice? Every year, the Telekom Design Gallery, a large, circular 16,000 sq ft space on the second floor of Deutsche Telekom's headquarters in Bonn, hosts more than 6,000 guests. The visitors represent a vast spectrum of industries and are experts in their fields. Politicians stop by. So do journalists, scientists, strategists, retail experts and, every so often, CEOs of giant international corporations. As soon as they arrive they are invited to interact with pieces of innovation they have never seen before, and often the equipment sparks lively conversation.

A progressive data service might encourage the politician to debate imminent tech regulation. A novel retail expression might urge a shoe manufacturer to consider the impending impact of digitisation on the factory floor. Eventually the discussions evolve into new ideas, some of which are tested live, there and then. Telekom's designers become collaborators with their guests, many of whom will leave the space with their mindset shifted and an entirely different outlook

on the industry in which they work. Almost all of the conversations will have revolved around a simple but critical question: What is the future experience of customers? More specifically: How does the digital future feel?

Answers can be found by experiencing hundreds of prototypes, many of which tweak cutting-edge technology to offer a snapshot of future use. All of them are hypothetical, to such a degree that often, when we first invite guests to engage with these prototypes, neither they nor us are able to fully appreciate their potential impact.

There are more than 250 use cases on display. Visitors can sit in a technological gaming chair that allows them to experience a virtual race with a virtual drone around the Neuschwanstein castle. Or they can participate in a virtual pop concert, experiencing the event as if they were the singer on stage. Or they can support a rocket engineer in the assembly of a FalconX rocket. All three of these examples use cutting-edge, augmented reality technology, developed for and within the gallery.

The prototypes populate three distinct sections of the gallery, each of which relates to a core area of daily human life and routine. One area introduces guests to technology that might affect how we work. Another area gives visitors a glimpse into the future home. A third area, called "Move", investigates travel and features a large representation of an autonomous vehicle. Visitors are offered tailored tours depending on their specific interests and industries, but given the section breakdown – home, travel, work – nearly everything on show applies to each and every one of us. Everything is interactive. Everything is experienceable. In many ways, the gallery is the closest we'll ever come to visiting the future.

The Design Gallery should not be confused with a showroom. It is designed to invite speculation, to convert visitors into change agents for potential futures. In this way, the prototypes are catalysts for defining business experiences of tomorrow. It's about enabling partners, scientists, designers and experts from all industries to articulate the corporate vision and make the strategy tangible. All for the benefit of our customers.

When a guest can see, touch and interact with a prototype, discussion is focused and streamlined. Of course, it is quicker and cheaper to create a presentation on screen than to build an environment that allows visitors to experience tangible aspects of the future. But doing it this way, visitors can dive into the digitally connected world of the future and discover it through the eyes of customers. By making the future tangible, it facilitates a platform that allows colleagues and guests alike to have in-depth discussions about what our future product and service offerings should be. Therefore, the Design Gallery becomes the innovation collaboration and engagement platform for the entire organisation that is on the path to a digital future.

The process can be slow, laborious and expensive. A gallery on this scale demands significant investment in both time and money. But often, seeing is believing. When you see, touch and hear something – if when introduced to a concept all of a visitor's senses are engaged – an idea becomes real. The physical space creates a sense of urgency and excitement. It sparks the imagination and nurtures understanding, which then creates opportunity, for both parties. And it allows us all to peek into the future. That's crucial. It's where we're headed, after all.

THE DESIGN
GALLERY

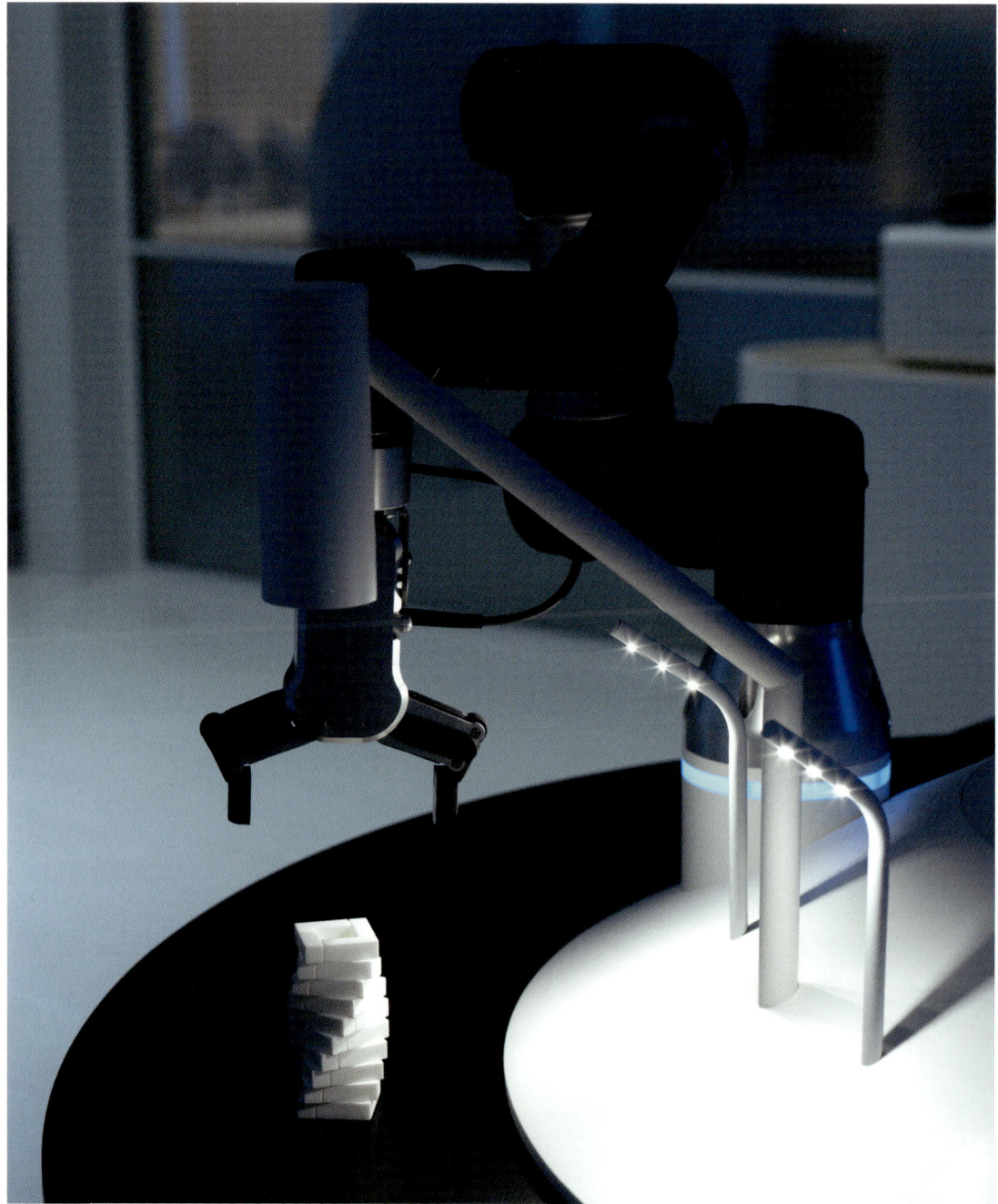

Predicting the future is the foundation for strategy work. In the contemporary climate, the exponential acceleration of technology means making predictions is a more difficult task to complete. Therefore the solution needed is for people such as CEOs, business unit leaders and innovation managers to create a set of potential futures as input for their planning work.

The biggest challenge is to identify unknown needs, ones that will emerge in the future and ones that are still latent. Users cannot yet express them, so the innovation manager must identify them. To do that, he needs to apply multiple strategies in parallel, hoping that one of them allows him to uncover the combination of an unmet need with its yet unknown technical solution.

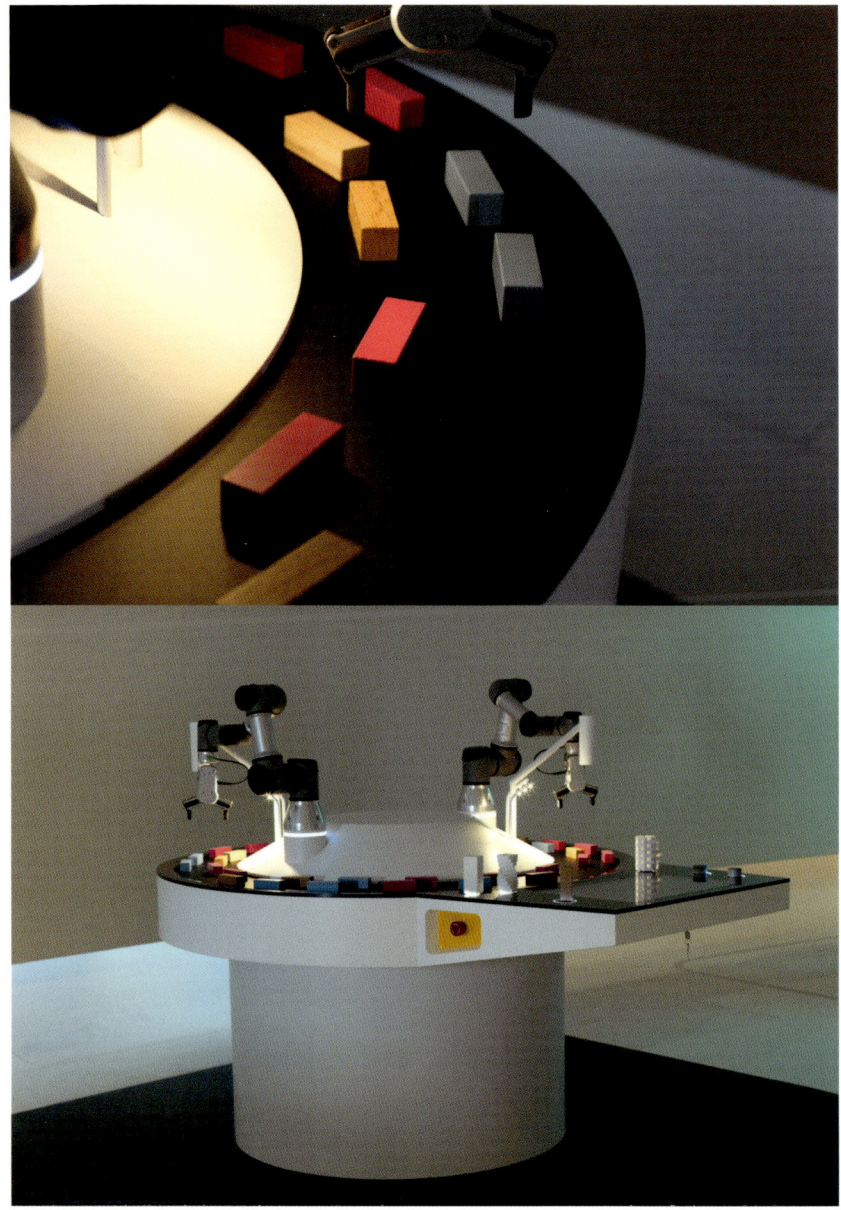

Creating the Telekom Design Gallery allowed us to expose the innovations that the organisation has been working on and developing. In this arena, clients in fact became co-producers of customised technologies and innovations. The Design Gallery has the effect of making room for the future because of its ability to serve as a venue for bringing the customer experiences of tomorrow to life.

Telekom's designers become collaborators with their guests, many of whom will leave the space with their mindset shifted and an entirely different outlook on the industry in which they work. Almost all of the conversations will have revolved around a simple but critical question: What is the future experience of customers? More specifically: How does the digital future feel?

Answers can be found by experiencing hundreds of prototypes, many of which tweak cutting-edge technology to offer a snapshot of future use. All of them are hypothetical, to such a degree that often, when we first invite guests to engage with these prototypes, neither they nor we are able to fully appreciate their potential impact.

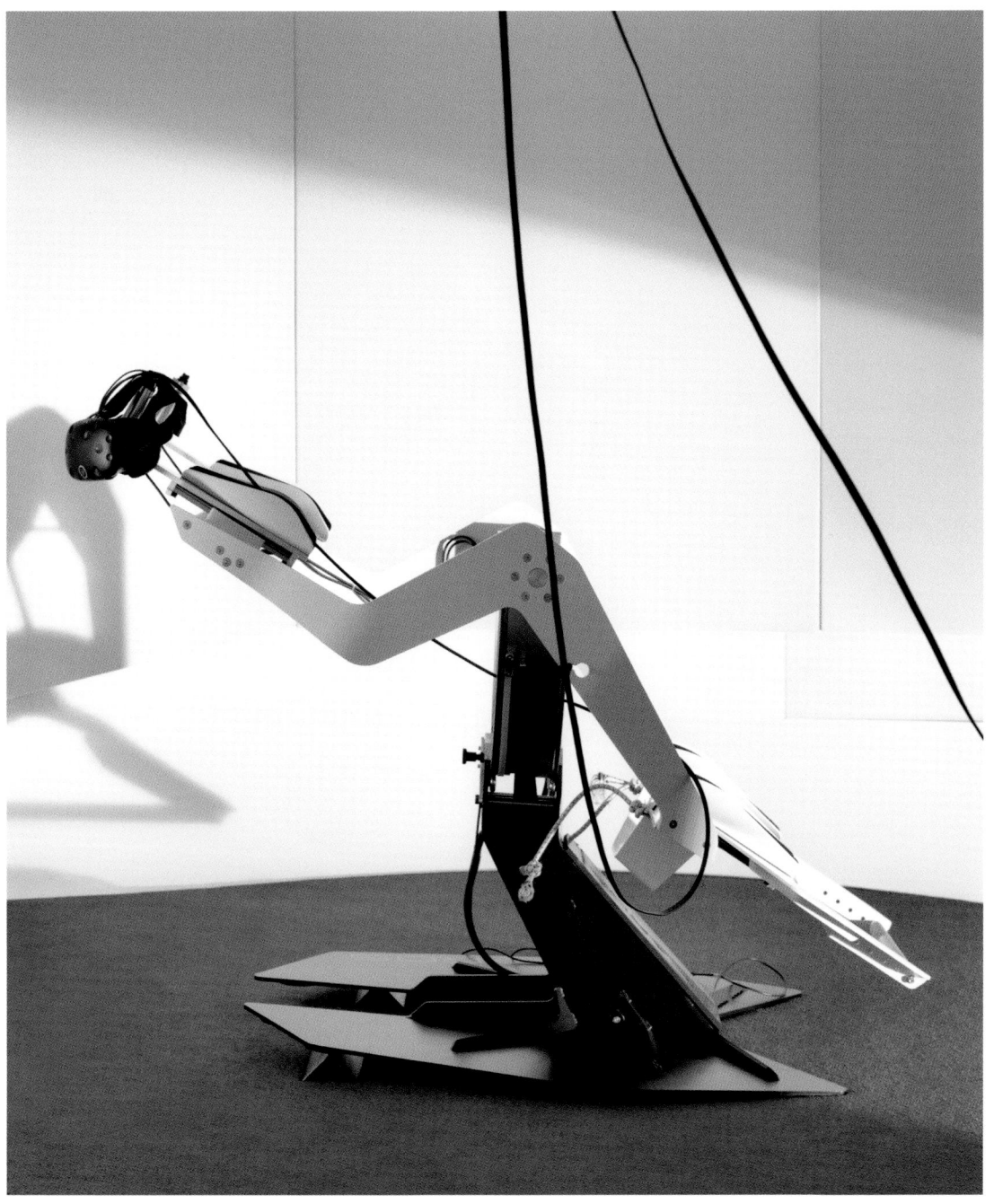

The prototypes populate three distinct sections of the gallery, each of which relates to a core area of daily human life and routine. One area introduces guests to technology that might affect how we work. Another area gives visitors a glimpse into the future home. A third area, called "Move", investigates travel and features a large representation of an autonomous vehicle.

The Design Gallery should not be confused with a showroom. It is designed to invite speculation, to convert visitors into change agents for potential futures. In this way, the prototypes are catalysts for defining business experiences of tomorrow. It's about enabling partners, scientists, designers and experts from all industries to articulate the corporate vision and make the strategy tangible.

Use design expertise to make scenarios accessible for laymen, and make them as tangible as a real experience for the customer, employee and top management. The Design Gallery therefore becomes the innovation, collaboration and engagement platform for the entire organisation on the path to a digital future.

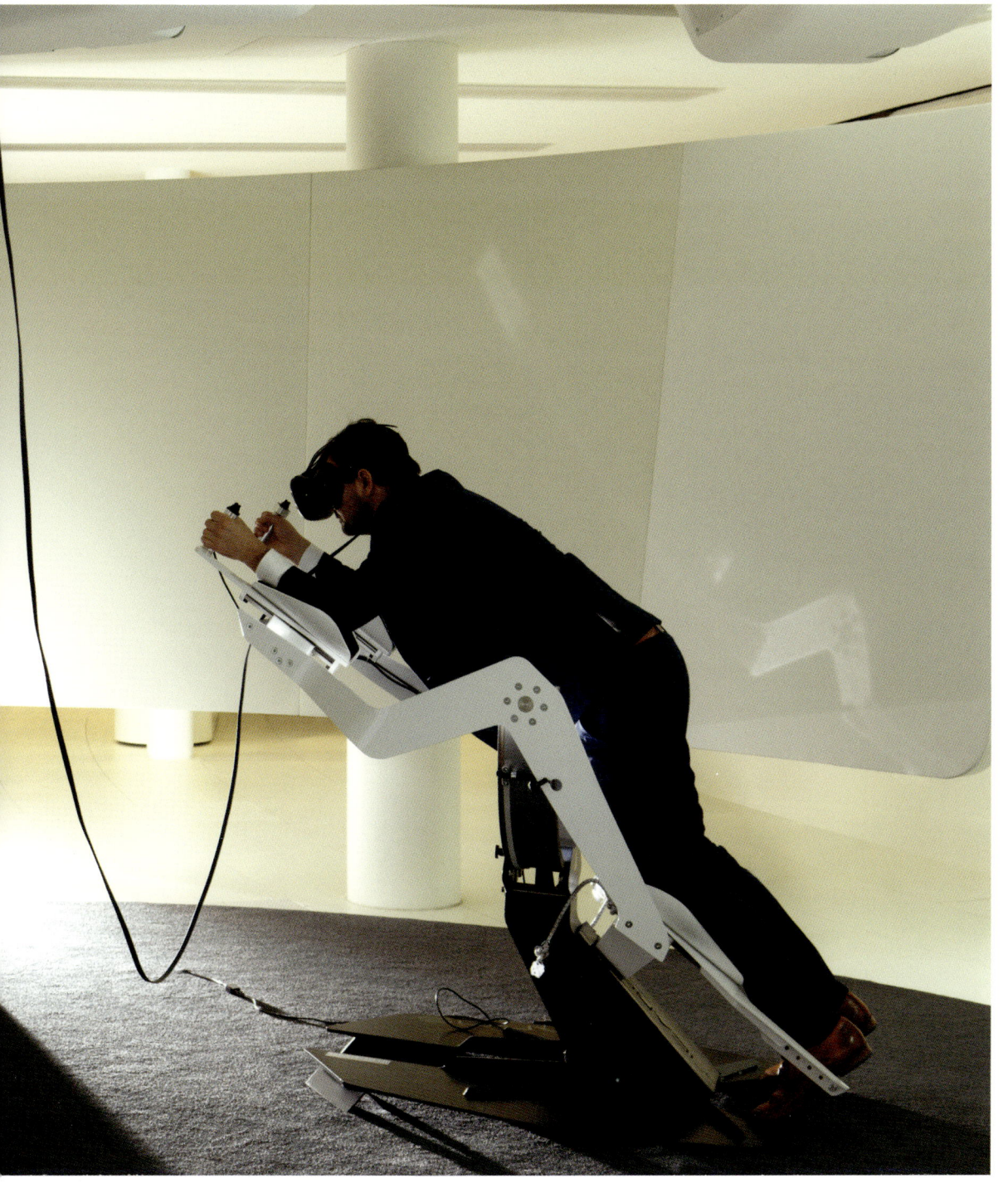

CHAPTER 3

EXPANDING DESIGN FROM PRODUCT TO STRATEGY

Designers unearth customer needs and anticipate scenarios. They translate them into strategies, products, services and meaningful experiences. Instead of merely beautifying nearly completed products, designers are now an indispensable strategic asset. Design drives systematic management of customer experiences and the complex innovation processes within large organisations. Building engagement through creativity bridges the corporate silos. In this way, design transcends its traditional domain of product and service design and becomes part and parcel of corporate strategy. A new role gets created: the strategic designer, a key figure for innovation and digital transformation. He mediates between technology and the living and changing environment.

Top actions for the CEO:

1. Set up external and internal customer intelligence. Because large organisations often confuse internal conversations about real customer needs and real insights.

2. Apply design's unique creative capabilities at the top level of corporate strategy, process and customer experience. Unlike any classic consultancy, design has the empirical and creative fantasy to predict the future.

3. Create win-win situations in your organisation to solve the friction between the horizontal needs of the customers and the vertical structure of the business units.

CASE STUDY:
DESIGN-DRIVEN
TRANSFORMATION

How does a group the size of Telekom manage to think about technology and infrastructure while also establishing design principles throughout the organisation – all so that they can sustainably improve the customer experience?

Deutsche Telekom finds itself in the middle of a spectacular transformation. It is moving from being a connectivity infrastructure company to becoming a front-runner of digitalisation. The company has set an objective to become a companion to its customers as they go forth into an increasingly complex, digital world.

The design of products and services was the focus of Telekom´s design function a few years ago. What was then a small team of designers has grown to become one of the largest digital design divisions within a German company. Meanwhile, over Twitter, a hundred internal designers were working collaboratively as part of a large network of freelancers, and domestic and international design agencies.

Consistency is essential for any brand. The look and feel of its products and services, as well as the way those services behave, have to be in harmony, otherwise user experience is grating and disorientating. That's why the design team combined design guidelines with a number of important digital assets within what we call an "experience toolbox", which ensures that divisions around the organisation can create a better and more seamless customer experience in more than 20 countries. To create this positive customer experience, a seamless "end-to-end" communication and cooperation between all services and products is of vital importance. To coordinate infrastructure and technology smoothly and provide a customer experience that is simple yet effective is a monumentally complex task. Especially for a large provider such as Deutsche Telekom.

In order to realise superior customer experience, the designer at Telekom, just as in any other company, has a key function: they take the side of the customer, not the organisation. The designer works on the customer's behalf, not of the company. The designer is always working to create the experience they know customers will want, and the products and services the customer desires.

Designers must observe, investigate and analyse their customers intensely. Therefore, customers' life characteristics, and most notable character traits, needs and desires have been systematically collected at the Telekom Design Customer Lab, then grouped and assigned to archetypal "Personas" representing different consumer groups.

This systematic customer segmentation allowed us to define different customer needs and expectations. It also helped us to select the most suitable candidates for focus groups in the Telekom Design Product Clinic. The focus groups were invited to the Customer Lab to evaluate new product ideas and prototypes and we integrated their unfiltered and immediate customer feedback into the design and development process.

To significantly improve the customer experience of products and services, designers have to think about what people really want, why they want it, what moves them and what they desire most. We want to turn this knowledge into solutions that may be invisible but unquestionably make life easier. We believe that humanising technology has a positive impact on the world around us. The design derives from empathy.

Listening to what customers really want is only a first step in the development of valuable solutions. Designers at Telekom have developed a strategic process of Design Thinking that is not only being put into action in their own units but has also made available to all international colleagues of the Telekom group an entire framework of methods and tools. As a consequence Design Thinking is not a specialised discipline only for designers. It fuses with the entire company strategy. Because every employee has an immediate responsibility for the end customer experience. There is no value in employing "Design Thinking" without actually "Doing". In summary, this transformed Telekom – establishing design and digital transformation abilities everywhere in the group.

"Design Thinking Doing" is the name of a handbook provided by the design unit to all groups at Telekom. It includes all common Design Thinking methods, processes and tools that have been adapted for standardised product development processes at Telekom. A digital version is available for all employees. Furthermore, the Design Academy was founded to help establish essential principles and make them clear to employees across all areas of the group. Telekom colleagues can take part in personal training, seminars and workshops here, to learn about best practice with the methods and tools we're providing. Already, more than 8,000 employees have been trained, all in the space of a single year.

ESSAY:
THE NEW ROLE OF DESIGN

D esign is important. It determines the relationship between us and the world we live in: what we wear, how we communicate, what we sit on, how our cities are laid out and how infrastructure works. To design means to shape the environment. Good design is behind every successful commercial product, and many designers have left a mark and sometimes an immortal legacy on the world. Raymond Loewy sketched trains, Frank Lloyd Wright gave us houses, Charles Eames gave us chairs and Coco Chanel gave the world Haute Couture. Dieter Rams designed hi-fi devices that are currently experiencing a renaissance as a backlash to a new and digitalised world. The success of Apple has been built on a singular design that starts with user experience and radically forces technological possibilities to fulfil its requirements.

Today, the role of design in product development has changed. It is now indispensable in the systematic management of customer experience and complex innovation processes within large organisations that still mainly work in isolated silos. Design thinking has become an integral part of corporate strategy. It is not only about product design anymore, but about the creation of complete business ecosystems. Within companies, the strategic designer is now becoming a key figure in the digital innovation process. His new role is to be a mediator between technology and the living environment. Due to the introduction of new digital technologies, the design of integrated hardware/software/service experiences is at the heart of every innovation process.

Here's an example. The introduction of self-driving cars requires a new kind of collaboration between automobile manufacturers and technology providers, municipal and governmental regulators, as well as innovative mobility providers such as car-sharing services, payment services and telematics. Their introduction also requires an extensive change in attitude and the dismantling of serious reservations in the mind of the end user. How will insurance companies collaborate with manufacturers and users to analyse risks? How will the data collected from self-driving cars be shared to control traffic, while at the same time be kept private, so that the car owners and passengers' personal data remains protected?

When creating a completely new business ecosystem, start-ups have the chance to set up a company that consistently adheres to design principles. That's relatively easy. But using the example of the automobile industry illustrates that large companies with origins in traditional industrial economies will find it more difficult to adapt, although the revolution in the automobile sector cannot happen without them.

For large companies the innovation process is getting more and more complex. But designers can reduce complexity, because they know customer needs. Design-driven innovators are able to anticipate scenarios for applications and to translate these into prototypes. These prototypes are then being further developed in interactive processes together with customers and technological producers. Such procedures of collaboration and co-creation force the designer and his work into a participatory, democratic framework. Therefore today, the designer does not democratise the design but the design democratises the designer.

The significance of the designer is still unbroken. But the heroic age of the designer is coming to an end. They are no longer the singular genius creating physical objects that stick around seemingly for eternity. Instead the designer is optimising the next beta version with their team. The designer's brilliance only shines for a second of certainty in an otherwise more fluid environment. Digital products, brands and services are never really completed. Therefore designers should evolve their role in the digital age.

In an increasingly complex world the consumer needs a powerful lobbyist on his side. To make truly worthwhile products and services, in every customer-oriented company, the designer should be fighting for the consumers and their best customer experience. The designer becomes the agent of trust for consumers. And design's role changes from product/service design to create meaning in the digital age. Making digital technology accessible for humans. And finally creating a landing point, a safe space, an anchor in the accelerating world we all live in.

HOW TO EXPAND THE ROLE OF DESIGN

In large corporations, design is usually an afterthought. Only when products and services have been defined (and, often, already developed) does the design team get the call to step in and "make it look good".

In the digital age, this process needs to be turned on its head. Design, jointly with product management, must lead the definition of product and service strategy, and the creation of the end-to-end customer journey with all subsequent levels of details. The design discipline in general and the design function in the corporare environment has to take a leadership role in orchestrating the company's and brand experience with all its touchpoints and act as the customers' advocate.

Actions:

1. **Hire/contract a strategic design team.**

2. **CEO promoting the strategic role of design – by visibly driving strategy creation jointly with design. And by participating in the end-to-end journey of at least one product/service.**

3. **Plan and execute internal stakeholder engagement from the inception of the strategic design team. Carefully plan win-wins. So that they see it as their idea and team.**

4. **Evangelise design across the whole organisation. Make design a mandatory participant in strategic and operational activities. From design as a department to design as an organisational capacity.**

5. **Set up outside in customer intelligence. Take customer centricity seriously. Because large organisations often confuse internal conversations about customer needs with real insights.**

6. **Train people across the organisation. With the goal to strengthen customer centricity and innovation capability by embedding design thinking in the organisational DNA.**

7. **Stage early wins. Apply design's unique creative capabilities at the top levels of corporate strategy, processes and customer experiences. Unlike any classic consultancy, design has the empirical and creative fantasy to predict the future.**

8. **Make the value of design as a strategic asset experienceable for everybody in the organisation. Build to test new ideas and hypotheses by making them experienceable by humans. This is crucial. It is extremely difficult to analyse the efficacy of a potential product or service by reading about it in a PowerPoint presentation.**

9. **Create win-wins to solve the friction between the horizontal needs of the customers and the vertical needs of the business owners.**

CHAPTER 4

HUMANISING ARTIFICIAL INTELLIGENCE

Humanising Artificial Intelligence – adapting it in order that it subscribes to the social, cultural and moral standards of contemporary society – is necessary for it to gain long-term acceptance among the wider society. This translates into thinking radically from a human perspective. The digital transformation of the economy, its business models and products, will serve as the foundation for a new, consistently successful global prosperity paradigm.

Top actions for the CEO:

1. **Choose a low-risk area and use case for the first release of AI with a use case that provides exponential returns. Exponential returns meaning that for a small investment you get a 10X return.**

2. **Leverage the exponential returns to spur an internal competition on implementing AI.**

3. **After the initial success build an AI deployment roadmap jointly with the key stakeholders. Keep in mind to make AI at least usable, flexible, versatile and modifiable by humans. Details on how in the following chapters.**

CASE STUDY: USING AI IN CUSTOMER SERVICE

Imagine, for a moment, that you have a problem with your router. It is not a difficult scenario to envision – routers can be complicated. Maybe a light is flashing that shouldn't be. Or perhaps the light has turned red instead of yellow-green. Or what if the lights have turned off altogether, and your whole network has shut down? You can't send an important email (it's too big to send using mobile data) and your children complain. What do you do? Do you turn the router off and on again, like everyone else does? Of course! You try that, but it's still not working. So what next? Do you actually have to call someone?

You know as soon as you ask that question what the answer is, and you don't like it: yes, you have to call someone. The realisation makes you grimace. You normally do everything within your power to avoid this kind of scenario. You had to call a hotline when you first set the router up and the experience was exasperating. The first number you dialled was incorrect. So was the second one. And you got cut off during the third call, despite having reached the correct department. The entire experience lasted more than an hour. You vowed never to call the hotline again.

But here you are with a broken router, and there's no avoiding it. You dial the number on the back of the product. An assistant in a department that does not deal with routers picks up the phone, realises after you've explained your issue that you've been put through to the wrong person and reroutes you somewhere else. You hear saccharine on-hold music – the same song over and over again. You're not even sure if the assistant in the next department will be able to help you, but you stay on the line anyway, because what choice do you have? The minutes ebb away. The cost of the call rises. And you are still on the line with no one on the other end.

Eventually you reach an assistant who can help. By this point, you're 25 minutes into your call. The assistant is friendly, perhaps because they can hear the frustration in your voice. When you explain the situation, the assistant tells you that the problem you're facing is common. He directs you to a page on his company's website that presents a list of router troubleshooting options. You access it with your phone, using mobile data. There you find an FAQ section. The issue you're trying to fix is at the top of the list.

It seems that many people have faced the same problem. The solution is simple. You carry out the steps. Ten minutes later you're back online. The email you needed to send is fired off to its recipient, and your children, who now have access to their cartoons, are content again. But it's taken 40 minutes to get to this point – too long, in your opinion – and throughout the experience you've repeatedly mouthed the question: Isn't there a better way?

*

While we were at Telekom, our team repeatedly asked the same question. For two years, while we were looking into ways in which we could improve the Telekom customer experience, we'd been made aware of reports of angry calls from customers. Some had contacted Telekom with an issue like the problem just described. Others had called about their bills or their mobile phone service or because they couldn't successfully install a Telekom product. Often the customer had reached the point at which he felt the only way to fix the problem was to call a Telekom helpline, but in many cases, those calls proved troublesome. Sometimes the issues were not dealt with quickly. Customers were rerouted from department to department, clogging up lines and leading callers to become frustrated. That incited a negative feeling towards the Telekom brand (even if a customer had a positive Telekom experience elsewhere – in-store for example). This also increased line and call centre costs for the organisation. Nobody seemed to be benefitting from the process.

The most frustrating thing of all was this: almost all the issues that customers called about were simple to fix. They were rudimentary problems with rudimentary solutions. And for the most part those solutions were available on the Telekom website, if only customers knew how to access them.

Our team began to gather a pool of concepts. They were ideas we could use to streamline the troubleshooting process. We knew that if the solutions to most customer problems existed on the Telekom website, then all we needed to do was direct customers online. But we also knew that websites can be difficult to navigate. It would be no good to encourage customers to look for help if, once they reached the Telekom homepage, they had no idea where to go next. Quickly an idea emerged: what if a service on the Telekom homepage could direct customers to the answers they needed? What if the customer could ask that service a question, and the service could answer right back, helpfully and quickly? Wouldn't that iron out most of the problems? All that negative feeling? All that money wasted?

The team began to test concepts and after a short time their experiments all pointed to the same conclusion: what was necessary was a kind of chat bot – a digital assistant that mimicked and expedited the function of a traditional call agent. The chat bot would exist on the Telekom homepage. As soon as a customer landed, the assistant would engage in a text conversation. It would be able to answer simple questions instantly. In fact, the more questions a customer asked, the more the chat bot learnt and the better it could direct them to the relevant solution, thus fixing the customer's problem. It would be both quicker and simpler than calling an agent on the telephone. And it would be much cheaper too.

We knew the assistant had to function seamlessly and we knew it needed to be precise. What was necessary was a kind of Artificial Intelligence. Soon we began to compare all available AI frameworks in the market in the hope of finding the perfect base platform for us to use. Twelve made it to the shortlist, which was eventually shortlisted to just one. Then development began in earnest.

*

Within two months, our team had created a digital assistant capable of answering rudimentary customer service questions. The technology was named Tinka. It was humanised, assigned a friendly female avatar, and added to the homepage of Deutsche Telekom's Austria site, where it began to politely answer customer problems. Within a month the team began to record its positive impact. Suddenly, customers could ask simple questions and immediately receive helpful answers without having to endure a time-consuming phone call. Our team watched Tinka's process closely, adding modifications and updating its dialogue capabilities with new language so the bot could answer more and more questions in more and more ways. Soon it could provide up to 1,200 different responses – all relevant to specific problems – all of which were created by our team. Questions to do with bills, products and services were suddenly dealt with cheaply and efficiently. Customers perceived it as caring, reliable, secure and sympathetic to their needs.

Before long Tinka was answering 120,000 questions from 50,000 customers and prospective customers per month. Consider this for a second: that's up to 50,000 less calls to a helpline every month, 50,000 less opportunities for a customer to become frustrated with the Telekom brand while being diverted from one department to another. The number of calls to helplines decreased monumentally. Organisational costs paid for by Telekom – line costs, for example – reduced. (The total cost of service decreased by 4%!) And positive feeling around the Telekom brand increased. The Net Promoter Score, a measure by which we can gauge the loyalty of our customer relationship, swung from -36% to +17% after Tinka's release. That is a phenomenal upturn.

When Tinka didn't have an answer to a complicated problem, the system referred customers to human colleagues, a two-fold process that seamlessly mixed AI with human capabilities. (Every conversation between a customer and Tinka is recorded and shared with a human advisor when the problem is handed over to them, in order that the employee might more efficiently decipher the right course of action.)

In an unusual twist, Tinka began to help human advisors, many of whom referred to the chat bot's expanding storage of data while speaking with customers. In this way, Tinka assumed the position of a giant knowledge database available for use whenever colleagues need updated information. The system became an expert in solving customer issues. In many instances, human advisors began to refer customers back to Tinka if they knew the system had an adequate answer, because the language the chatbot uses, and the animations it provides to customers when they need to troubleshoot an issue, offered more clarity than the advisor could.

The digital assistant was such a success in Austria that soon our teams began working on an updated version – added intelligence, added personality – for the German market, which is much larger and far more complicated regarding the products and services it offers. The system is now able to develop and construct its own answers to new questions.

It has learnt to trawl the internet for new information and terminology related to customer issues. It is also able to provide customers with details of newly launched Telekom products before our human advisors can. The system is being constantly updated. It is regularly equipped with new dialogue, more streamlined functionality and greater intent recognition – the ability to understand the customer's intent. This ability is a core functionality of the AI as there are many different ways to express the same intent. As an example, you might say: "Turn the light on" or say, "I want it brighter". In both cases the intent is the same – lights up. However, it's expressed very differently. In the meantime, the system has evolved and is able to identify and understand quirks of language. It can be logical. It is able to read a customer's emotion.

And it will only become more intelligent. According to a German survey, four out of every ten people believe that services like Tinka are important when it comes to handling orders and complaints. That percentage will only rise as systems become more intelligent and subsequently more helpful. This will not stop at chat bots.

Our teams are now also developing voice recognition software – systems that will be able to speak and reason with customers. It will be able to recognise the emotion in customers' voices and the different kinds of language – regional slang, for example. It will understand when a caller is working within a time constraint and sometime soon it has the potential to develop personal and longstanding relationships with customers.

No longer will a customer be forced to deal with several different advisors just to sort out one problem. They will not be required to repeat their issue again and again. The customer experience of the near future will be consistent and enjoyable. Conversations will begin where they left off. The system will have access to swathes of data, both about the individual caller and about Telekom itself, which will expedite the troubleshooting process. Above all else, the system will help more and more people, just like it has done in Austria.

ESSAY:
A HUMAN AI

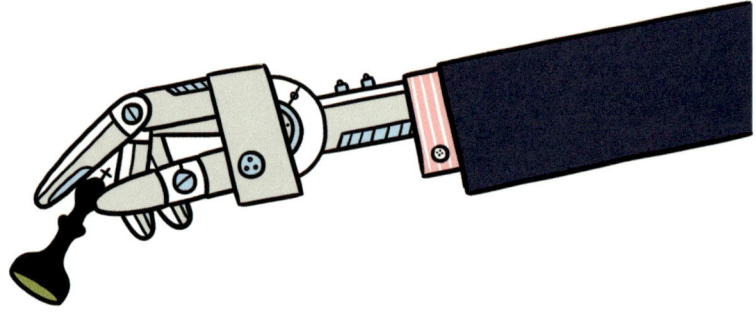

In 2017 a computer program called AlphaZero beat the existing chess world-champion computer programs. AlphaZero was created by Deep-Mind, an Artificial Intelligence research company owned by Alphabet, Google's parent company. The program was launched not long before it was due to play its chess tournaments. Within 24 hours it had achieved an unparalleled level of strategic ability. DeepMind's engineers programmed AlphaZero to train via a process of "self-play". In other words, the program learnt by playing games against itself and becoming its own teacher. It knew nothing about chess when it was created. Because, unlike earlier forms of AI, it did not rely on human expertise to learn the basics of each game; its strategies were novel, creative and surprising. With every new game it played, it tested a series of complex potential moves, quickly raising its level of ability. And it did this much faster than any human could hope to do.

Whether we like it or not, AI is here to stay. AlphaZero is one indicator of its enormous power in the world today. It is a form of intelligence that learns from and by itself. It displays ingenuity and creativity – traditionally human traits. But this is only the beginning. In the new world of AI, every object imaginable will display superhuman intelligence.

Humans are often driven by convenience: if a person can be lazy and someone else does things for them, then they will choose that option. With the rising capabilities of AI, we might hand over more and more activities and work to the AI. This might lead us to surrender control of great portions of our lives. What happens if AI begins to make decisions that have a negative effect on others? How can we hope to maintain autonomy without full comprehension of the reasoning behind the choices we make?

But AI also creates opportunity. Imagine the value an organisation can create by developing an AI system that is usable and understandable by humans – that enables us to make better decisions as individuals, and as a species.

Many of us interact with AI every day, and the integration is becoming more and more entrenched. It's a topic that concerned the late Stephen Hawking, who said, "One can imagine such technology outsmarting financial markets, out-inventing human researchers, out-manipulating human leaders, and developing weapons we cannot even understand."

Hawking also voiced concern. "Whereas the short-term impact of AI depends on who controls it, the long-term impact depends on whether it can be controlled at all. All of us – not just scientists, industrialists and generals – should ask ourselves what we can do to improve the chances of reaping the benefits and avoiding the risks."

What can we do? One thing is to begin to create Artificial Intelligence systems that act on our behalf and are understood by humans. That means is one way of humanising AI. Humans connect through emotional, empathic interactions; empathy is the glue of our society. But AI has not yet been programmed to comprehend or share human feelings. It does not showcase sympathy. It does not understand human motives and drive. It isn't aware of the things that really concern us, or the things that make us tick. To become beneficial to society, AI needs to learn all the above. AI must become civilised to bring it to social, cultural and moral levels that correspond to current social norms. If it does, it will become a trusted partner that can enhance who we are.

Why is this important? Because Artificial Intelligence has the potential to develop into a dark force. As we each evolve into collections of data, the companies that control that data have grown rich and powerful. These organ-isations know things about us that even we ourselves don't really know. They use algorithms to find patterns in our data to better recommend products and services to us. But that's not all. With all that data, what will happen to our privacy? What if battlefield AIs make catastrophic decisions without human oversight? And a much larger question: what happens if AI's goals begin to differ from Man's? Humanising AI reduces those risks. It makes AI an extension of human decision-making, not a threat to it. A tool, not a weapon.

Let's take work as an example. AI is already performing certain mundane tasks better than humans, and it has the potential to be applied across all areas of industry. Customer service departments. Magistrate courts. Assembly lines.

Reception desks. Doctors' surgeries. Supermarket aisles. Newsrooms. Soon it will appear that AI is able to perform all tasks better than humans. AI will be more reliable, more productive, quicker and less prone to error. Additionally, AI will never have to stop to eat, sleep, or gossip around the water cooler.

There's no doubt AI will soon oust some of us from our jobs. But what if AI can be trained to help humans in new roles, rather than replacing us altogether? What if AI helps us to become better at our jobs? What if we allow AI to perform the prosaic elements of our working processes, leaving humans to excel in roles that demand creativity, empathy or a capacity for meaningful relationships?

This humanised version of AI is better for society and better for business. It delivers better results. It considers the people it affects. It aligns with the moral compass to which society subscribes. It is more easily trusted.

How do you humanise AI? Think of how parents raise children. We want our offspring to act morally and ethically and we follow parenting methods that encourage that to happen. Organisations must do the same in the case of AI. They must breed AI systems that learn the societal norms of responsibility, fairness, compassion and transparency. This isn't happening now. Most enterprises consider AI a tool. They use it to find patterns in data for commercial gain. We don't expect AI to explain its decisions. We don't expect AI to make concessions to collaborate more effectively with humans, even if the decision results in lost profit. We don't expect it to consider its effect on us, the impact it has on society. But we really should.

The sooner we create a responsible AI, the better it will serve our needs and the better we'll be able to use it effectively. Any existing conflict between AI and society will quickly dissipate. It will turn a potentially dangerous technology into a powerful human collaborator – into a personal tool. And everyone will win.

HOW TO HUMANISE AI

AI needs to grow up. We have to nurture it in a multistage process. This will result in five levels of maturity, marking the progress and humanisation of AI.

Actions:

1. **Start by introducing level 1 – adaptable AI. The technology must be usable, flexible, versatile and modifiable by humans.**

2. **Transition to agreeable AI. This means the technology will be pleasant, manageable and easily comprehended by humans.**

3. **Make AI docile. So that humans perceive it as gentle, meek, smooth, patient, tolerant and compliant.**

4. **Proceed to a compassionate AI. Shaped so that humans perceive it as caring, tender, warm-hearted, peaceful and sympathetic.**

5. **Domesticated. Artificial Intelligence must be domesticated so that humans are able to completely trust in it. Technology must be devoted, disciplined, reliable, risk-free, safe and secure.**

In the development process of the humanisation of Artificial Intelligence, at best, we currently find ourselves at stage two of the above model.

To give an example: Telekom has made significant investments in Artificial Intelligence systems. The Tinka project is one example, and it is helpful to refer to when considering how your organisation might begin to better use AI. Here's why.

First, our team rolled Tinka out in what is widely considered a smaller market interested in innovation and novelty. Due to this nature, Telekom Austria is highly successful and a good region to explore new approaches.

That made it a fertile playground on which to experiment with customer-focused AI. We could introduce Tinka to Austrian consumers and record the service's impact without having to worry too much about its effects. If it performed poorly, we would remove the service, and the impact would be limited. If it went well, fantastic!

Second, we ensured that Tinka was set up to benefit customers, rather than collect data from them. In this way, an Artificial Intelligence system became a trusted partner. When we spoke to users about Tinka, they considered the service overwhelmingly helpful. Nobody referred to the algorithm as "scary" or "dangerous". In effect, we had humanised AI.

As the leader of an organisation you have an opportunity to do something like what we did with Tinka. But we advise taking these two factors into account. First, identify a portion of your business in which you can experiment with AI without having to worry about causing irreparable damage to your business. Perhaps in one of your smaller, existing markets, or during the rollout of operations in a new one. The impact of your system will likely be positive.

Second, you must create an AI system that places consumers first. The AI you use must help users for it to be effective. Otherwise the algorithm will not be trusted, and you might face potentially harmful consequences. If a customer sees that an AI algorithm is working on their behalf, they will think of it positively. And that positive thinking will extend to your entire brand. This is what we mean when we say an organisation must view Artificial Intelligence as more than a tool that is used to collect data about your consumers. (Think about how people reacted to Facebook after the Cambridge Analytica scandal.) Any AI you use must be as transparent and honest as possible. Humanness is imperative.

The next steps are easy. As soon as you have tested your AI system in a small market and received positive feedback, you can roll the system out into other areas of your business. You might need to tweak the technology in places. Using the example of Tinka, the language it uses when speaking to consumers in Austria is different to the language it will use when speaking to customers in Germany. But the strong foundations have already been laid.

CHAPTER 5

CREATING PERSONAL INTELLIGENCE

Today, many people already interact with AI – for example, when Amazon's recommendation system proposes products for users to buy. However, that interaction is like watching TV, where stations control the content. Amazon controls the AI. And customers can merely select between results.

Let's imagine a different AI. One that is a personal tool for everybody. Tools are extensions of human capabilities. They are used by individuals. To build something, to create an impact, to change the world around us. Tools are controlled by the user. A great user experience will thus be key to making AI user friendly for everyone – to democratise it.

Countless technologies have undergone a process of democratisation. In many cases, like the computer, say, technologies were designed for use by research and government organisations, and only later were they adopted by large corporations, after which they eventually became available for personal use.

Similarly, Artificial Intelligence will transform into Personal Intelligence. From AI to PI. A smart tool for everyone. Personal Intelligence requires applying user experience to Artificial Intelligence in a radically new way and design will play a prominent role in this process.

Top actions for the CEO:

1. **Establish a PI SWOT team to kickstart understanding, including designers and AI software experts. Have them head OUT, ignoring your product managers who will tell you there is nothing new to find.**

2. **Head OUT internally: learn PI through internal use.**

3. **Choose simplifying complex manual interaction with legacy IT as an example. People hate them and will be amazed seeing this happen.**

4. **Head OUT externally: next, build a first PI tool for customers. Democratise hard by transforming AI to PI.**

5. **Head OUT strategically: disrupt or be disrupted. Map the 6 Ds of exponential technologies to your business – digitisation, deception, disruption, demonetisation, dematerialisation and democratisation.**

CASE STUDY: WHY WE NEED TO GET TO KNOW OUR CUSTOMERS

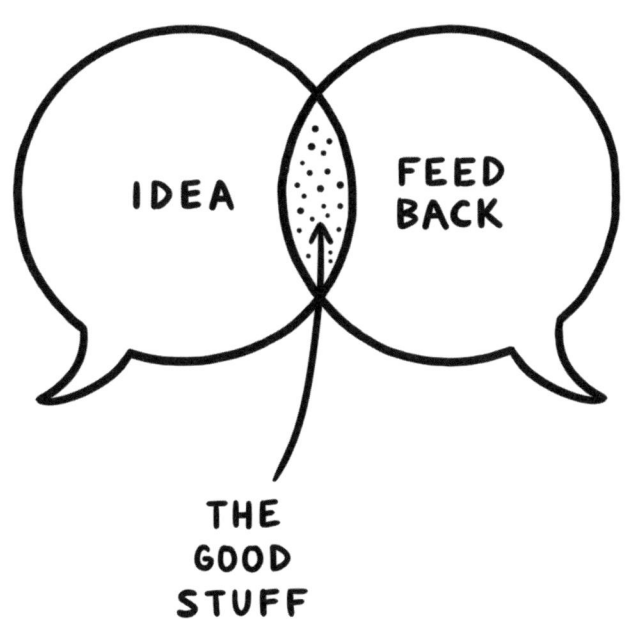

One day not long ago, a few members of the Telekom product and innovation team met in a meeting room at our headquarters to discuss the merits and potential pitfalls of a soon-to-launch product. Some in the room believed the product would be a big hit. Others questioned its relevance among its target demographic. Before long, a healthy debate broke out, and it raged throughout the morning. When questions about the product's efficacy surfaced, the designers attempted to answer them using limited existing data and customer research, but the information could only address certain issues and the questions persisted. Exasperated, the team struggled for answers. Then someone suggested they go for a walk.

What happened next transformed the way products and services are conceived at Deutsche Telekom and marked the beginning of a major cultural shift within the organisation. The team exited Telekom's HQ and headed out into the city, first to Bonn's busy Central Station, and then into the surrounding streets. They had with them a prototype of the disputed product, and soon they began to ask passersby for their opinions: how it looked, how it functioned, whether they would use it in their homes, if it was something they really required.

The team recorded responses longhand in notepads, scribbling furiously, underlining illuminating comments. The questions the team hadn't been able to answer back in the meeting room were now being answered by the kinds of people for whom the product was being created, and the insight was invaluable. By the end of the day, the team had collected much of the information it needed and, over the next weeks, the product evolved to better meet the needs of its intended customers, who had suddenly become directly involved in its development. Whenever a question about the product's functionality resurfaced, the team knew all it had to do was take a walk. Critical answers lay just beyond Telekom's front door.

In the past, direct customer feedback had played a minor role in the conception and development of products and services propositions at Deutsche Telekom due to its past as an infrastructure provider. This is a position that is not unique among telcos – or large organisations more broadly – which is shocking given how much insight it provides. During product development, Telekom's consumers had been consulted, but often those consultations took place late on in the design process, when resources and budget had already been deeply invested and the product was near-finished. At that point, a product's course could shift only so much. Sometimes it would launch to great effect. Other times, not so much. If customer reaction to a prototype was deemed wholly negative, a concept would be shelved, and the resources committed to it would be wasted. Often products launched to market that customers didn't need, and the entire project would end in failure. All because nobody thought to involve the customer in the early development process and in a very unpretentious way.

Not talking to the consumers for whom a company is creating products and services is counterproductive to the sound running of any organisation, so we decided to make changes. First, we internally published 16 customer personas using a mix of quantitative and qualitative data, and we referred to that data at every stage of a product's development process. We used these personas to frame the questions we asked consumers. And by doing so we were able to better get to know them.

Next, we installed in our teams the confidence required to be ruthless early in the product development cycle. If, when we presented an idea to one of our persona groups, it received a positive reaction, the idea would progress. If it didn't – even if reaction was muted, neither positive nor negative – the idea would be dismissed. This approach makes practical sense. Concepts with inherent fatal flaws could be identified and extinguished before running up major expense. But the greater change we made was this: whereas previously a concept's development relied on board members giving the go-ahead – a practice that is both wildly subjective and strangely traditional among organisations at which top-down hierarchy is respected – we requested that from now on a different group of people should have the final say: Telekom's customers.

Before long, customers began to play a pivotal role in the development of Telekom's products and services, often very early on in the process. The practice of approaching customers in the street and recording their responses to new ideas – experiences our team later called Service Safaris – became *de rigueur* during early concept development. Opinions would be collated, dissected and later used to guide the direction of ongoing concepts.

We began to map our customers' journeys. We interviewed consumers in our stores about their habits. We talked to people on public transport. We set up focus groups. We identified consumer pain points. We began to understand what went well when customers experienced our products and services and what went horribly wrong. In short, we learnt more about them: who they were, how they behaved. And we did so just by talking to them. All to gather small pieces of information that when collated, could have profound effects on the success of a concept in development.

Customer centricity – where the customer is central to the development of a product or service – has long featured in the product development processes of progressive organisations. There, employees understand that all innovation should centre on people's needs, that we need to involve customers in the decisions we make. The thinking goes that the better we understand a customer, the better we can create something that is helpful to them or desired by them, and the better a product performs on the market. Sounds simple, right? But large organisations around the world, often burdened by bureaucracy, have either failed in their efforts to implement customer centricity successfully, or simply haven't tried. It is now an approach that sets Telekom apart from its competitors.

The conversations really helped. Over time, our teams began to better understand the immediate and future requirements of the individuals for whom they were creating products. They learnt, for example, that some customers found it difficult to set up a new router – too many wires! – and that others were irritated by the number of streaming platforms they now subscribed to and hankered after a single-destination service that aggregated disparate services. As the number of conversations mounted, so did the number of problems or desired services customers shared. And as soon as we had that information, we could go on to design products and services accordingly. Our primary strategic driver became human requirement. The customer became central to everything we did.

*

Sometimes, though, it wasn't enough to approach people on the street. Soon, the design team began to invite groups of customers into our headquarters for involved sessions of product analysis, during which they would be presented with concepts and asked to comment. When they were available, early prototypes would be passed between customers, who would analyse the product through the prism of their everyday lives: Will I use this? Is this helpful? Is this something I don't already have? Because customers were regularly presented with ideas that had only recently been constructed, prototypes did not always exist. In those cases, attendees would listen to a verbal presentation, and a heated discussion would ensue. We called these sessions Customer Labs.

For our designers, Customer Labs, which fell under the auspices of an internal project the team called Panther and which later won the German Innovation Award in Gold, were thrilling. It was disappointing when an idea received negative feedback, but it also paved the way for new concepts, averted unnecessary toil and saved money. And when comments were positive, the design team could carry on with assurance. Every now and then a third response emerged: when consumer conversation brought to light ideas the design team had yet to explore, which, rather than end or progress a concept's life, might instead significantly alter its direction.

On one occasion a concept was rejected by the demographic for which it was designed but later adopted by an entirely different set of people, for entirely different reasons. The idea subsequently changed course. It became effective in a new capacity for a new customer group, and the work and budget that went into its early development was saved.

Very quickly the sessions became vital to the design process. No idea progressed until it had been analysed by a group. But the approach raised a few eyebrows internally, particularly when the balance of power began shifting from the organisation's board members – those who had previously held the final decision over whether or not a product would develop – to our customers, who would become the final users of our products and services.

During Customer Labs sessions, senior executives were requested to sit at the back of the room and observe, an astonishing demand given their high status. Here were some of the organisation's key executives being asked to kindly be quiet, to have no say on which products and services their organisation might release into the world. This didn't happen at other telcos. It rarely happened, period.

Partway through one session, during which several ideas were showcased to customers, we were asked to identify the concept we liked most. We replied that we would prefer not to say. In the context of the session our ideas mattered very little. The concept was not being created for a demographic of which we were a part, so why should our opinion matter? We didn't want to fall into the trap of impressing our own bias onto the idea. Instead, we replied that it is the people who will use the product who should make the decision about whether it goes ahead. In short: no matter what we or the board thought, the customers were right. And only through conversation with them could what was right be aired, documented and used.

Have the Customer Labs worked? They have. We've learnt, for example, that context is what matters most for our customers. Here's a specific example. Picture a consumer listening to music via a voice assistant. The consumer is listening to a Rolling Stones album, but the song being played is not one of their best and the consumer provides the assistant with a direction: "Play next track". As humans, we intuitively understand what this command means because we understand its context: the consumer is listening to a Rolling Stones album and he would like the next track on that album to be played. And yet all current voice assistants fail to understand. What to humans seems like a simple command is lost to the AI system. The voice assistant simply replies: "Don't understand."

Understanding that context is crucial when creating an effective voice assistance system was a huge learning curve. We applied it to Tinka, and we quickly realised that we would need to apply this to all of Telekom's conversation assistants. But without the Customer Labs – without hearing from consumers about what was good with the technology we were creating and what was bad, without hearing that Rolling Stones story – we never would have gained that knowledge.

*

In understated terms, our great learning was this: talking to customers helps, so long as the dialogues are thoughtfully structured. We're now wholly confident that the products and services we create at Telekom are of benefit to our consumers. Involving consumers in the product development process has saved us time and money – and plenty of sleepless nights.

But over the past few years we've also learned that customers expect all objects and services in life to behave intelligently. They expect products to adjust to their specific needs. It doesn't matter if it's a voice assistant or a smart fridge. Individual customers now demand that products react to their individual preferences.

That's a tricky realisation to reach, because it means the era of the demographic is coming to an end. Consider this: ten years ago, it was assumed that two 30-year-olds with similar jobs, backgrounds and interests were likely to be attracted to the same product and would use it in similar ways. This made creating products straightforward. If you understood one of the 30-year-olds, you also understood the other. By creating a product for one person you could create a product for everyone like them.

Now that's not the case. Large customer segments tend to no longer exist. Today, those two 30-year-olds might appear similar on paper, but they have wildly different product requirements. One might want the smart fridge to give them advice about their diet. The other might want the same product to focus on supplying information about when certain foods are about to go bad. The change is slight, but it's hugely significant. These two people require the same product to have different capabilities. And it's not just fridges. It's every product and service imaginable. Consumers are all beginning to demand functionality that is specific to their individual needs. That means we're all beginning to have less and less in common with other customers.

Now when we talk to customers we speak specifically about their individual needs and work hard to develop products and services that can be tailored to personal preferences. With the development of smart products, that's becoming easier. Artificial Intelligence, like the system behind Tinka, allows us to adapt user experience to the needs of one person, not a group of people. Functionality can be customised. Products and services can serve individual needs.

And all of this came from a walk.

ESSAY: INTRODUCING PERSONAL INTELLIGENCE

A rtificial Intelligence is a horizontal technology, which means it is applicable in many areas and industries for many different use cases. AI can now become a tool – usable by everybody. In addition, we must civilise Artificial Intelligence. And bring it to a level of social, cultural and moral development that corresponds to current social progress. In summary, this translates into thinking radically from a human perspective.

A humanised AI can then become the foundation for the digital transformation of the economy. As it impacts all business models and products, it can then serve as the foundation for a new, consistently successful global prosperity paradigm.

Artificial Intelligence must complete the transition to Personal Intelligence: from AI to PI. It must become a tool for everyone. A technology that helps humans instead of hindering them, that allows them to become sovereign agents and users that profit directly from technological progress. How can we get there? By applying methods derived from user experience to Artificial Intelligence.

If we equip Artificial Intelligence with an understanding of human motives, concerns and emotions, it begins to work on our behalf. It becomes helpful and meaningful to humans. It becomes a tool.

Tools have played a crucial role in the evolution of mankind. We have been using them for hundreds of thousands of years. And whenever a tool became accessible to everybody, it had a substantial impact on society.

Tools are typically extensions of human capabilities. A hammer is an extension of our ability to pound objects such as nails. It allows us to pound harder. A microscope is an extension of our ability to see. It allows us to witness objects in greater detail and with greater focus. And a pocket calculator is an extension of our ability to perform simple mathematics. It allows us to calculate larger numbers, faster.

Now think of the commonalities between these three objects. They all enhance an existing human ability, and each is progressively more difficult to use than the last. In the future, AI must be functional *and* usable. It must be a tool everyone can use easily, and we must all profit from it equally.

How can an organisation help achieve this? It can create a system that combines Artificial Intelligence with the best of user experience. We will call it Personal Intelligence. It is Artificial Intelligence that works from the human perspective. It makes Artificial Intelligence usable for everybody. Let's put it as a simple equation: AI + UX = PI.

Why is this transition so important? As disruptive technologies continue to overhaul the way we live, our personal preferences, needs and cravings are becoming increasingly individualised. In the past, it might have been true that the products I desired were the same objects that my friends also craved. A car I lusted after was also lusted after by people just like me – individuals with similar backgrounds, education, experiences and behaviours. The same was true for pianos, shoes and televisions. Manufacturers were able to group similar individuals into large cohorts and create advertisements that spoke successfully to broad customer segments. Today, that "one size fits all" approach is becoming more and more ineffective.

In our time, self-design has become the mass cultural. The internet is a place for self-presentation. From Facebook to Instagram, everyone is expected to be responsible for the image they present to others. Self-design creates a second, artificial representation of the self: a digital twin.

Soon, almost everyone will have a digital twin consisting of all their user experiences and interactions with digital services and products. The digital twin will represent the unique choices and tastes of every single user. As a result, there will be as many individualised models of one and the same brand. At that point customer segmentation will reduce to one.

Today, companies group customers into broad segments that have similar behaviour patterns. Tomorrow, companies will target customers as individual users and build experiences specific to them and their needs. The goal will be to deal with every user as a "segment of one," regardless of how, where and when they interact with the brand. Thanks to extremely flexible design, it will be possible to create an individual user interface for each person and his digital twin. Customised to their unique mental image of the brand, it will distinguish itself from those of all other users.

In the field of digital services and brand experiences, user experience benefits from Artificial Intelligence's economies of scale. In the personalisation process, data derived from a user's profile and real-time context will help create experiences generated specifically for them. It will even be possible to employ different design elements for different user experiences. Today we can calculate many millions of application scenarios, experiments and prototypes at a very high speed in order to truly configure smart applications for a specific person. And that means everybody will interact with AI.

Technology enables brands to interact with users in new ways, such as virtual assistants that can engage in a one-on-one conversation and deliver real-time, personalised service at great scale.

Personal Intelligence is entering the mainstream of customer experience, while leveraging information gathered in digital identity management and personalisation capabilities. The system will improve over time, becoming more intelligent, personalised and helpful the more we learn about the user. Personal Intelligence allows more proactive experiences for the user and those experiences will feel like personal relationships.

With Personal Intelligence, consumers can build their own representations of products and services. Personal Intelligence will boost self-design. There are many business models imaginable around Personal Intelligence: especially ones that enable the user to adopt and change the design of products. Personal Intelligence allows mass customisation. That means the production of highly individualised products under the cost-effective conditions of mass production. This will not only affect fashion, where mass customisation has been tested over a long period of time. In the future, everybody will be able to design a model of a product and print it out in a 3-D-Copy Shop around the corner. Everybody will be able to use Augmented Reality technologies to change the design of their environment, home or workplace. Personal Intelligence will provide real-time services for every location or situation. Finding the next exhibition or concert of an artist who is relevant to the person. Or calculating a trip from A to B.

Personal Intelligence for digital and telecom services will learn the individual behaviour patterns of the users. Personal Intelligence will help people to better understand their life. Connected to collective intelligence in the cloud, Personal Intelligence can predict behaviour patterns and can help to better synchronise social relationships. The more somebody knows about their partner or colleague, the more empathic they can be. The most important aspect is this: the data gathered and compiled by Personal Intelligence will stay with the user and will not be monopolised by big anonymous companies.

But there is even more. Personal Intelligence will comprehend the emotions of others, enabling it to tell us when the best moment might be to ask our boss for a raise. It will understand our design preferences and the way our moods work over the course of a typical day, so it will use Augmented Reality programs to personalise the interiors of our homes and cars to best suit our changing states of mind.

Personal Intelligence will have a copy of our complete biometric run-down. It will be able to connect with local manufacturers to create a 3D-printed version of a sneaker that both perfectly fits the shape of our feet and is best suited to the sports we prefer. PI will even suggest to us who we should hang out with, who we should avoid and with whom we should pursue a romantic relationship. It will work tirelessly to make our lives better. It will also inspire by supplying us with surprising and unexpected information. It will help us understand the decisions we make and why. It will help us better know ourselves.

And this system will further improve over time. Leveraging the loads of information it gathers, PI will become more intelligent. It will learn more about us and become more and more helpful. And most important of all: all the personal information it collates and shares will be vetted by us, the user, allowing us to maintain autonomy over our data. It will be shared with the people we like, the organisations and governments we trust and no one else. AI will be domesticated. Personal Intelligence will transform the individual from an object of technology into a subject of free will and make them the sovereign master of their very personal universe.

CHAPTER 6

PERSONAL INTELLIGENCE AND THE NEXT GENERATION INTERNET

CASE STUDY: WHAT DOES A NEXT GENERATION INTERNET MEAN?

In its brief history, the internet has evolved through multiple iterations. It began, in the 1970s, as the ARPANet, which collated divergent networks into a single network – a network of networks. The ARPANet was the technical foundation of the internet we know today, but it was never used by the public, and to most of us it would be unrecognisable. Funded by the United States Department of Defense, among other military institutions, it was used primarily for military research. ARPANet subscribers, most of them military men and women, could use the network to send personal messages to each other, but other forms of traffic were outlawed.

Two decades later, in the 1990s, a new internet emerged. This was the internet we know as the World Wide Web – an "information space" invented by the English engineer and computer scientist Tim Berners-Lee. Whereas the ARPANet was used by a limited number of private subscribers, the World Wide Web was created for global use. It introduced to the world the web browser and subsequently, web pages – hyperlinked text documents through which users could publish masses of information. Using this internet, large organisations began to create rudimentary websites onto which they placed information about their businesses and individuals began to construct rudimentary blogs, onto which they gainfully shared information about their lives. This is the internet that created Amazon, Yahoo and Google (as well as file-sharing sites like Napster). It paved the way for mass, global information-sharing and became central to the emergence of the Information Age. It changed our world.

But even that wasn't enough. Next came the app-based internet, in the 2000s. This internet emerged in tandem with the invention of mobile internet devices – the smartphone, for example. Suddenly web pages became less crucial to the sharing of information because businesses could now also create web applications – computer programs designed to run on mobile devices – which were available to download, often for a fee, from digital stores.

The app-based internet helped to usher in the age of social media – Facebook, Twitter and Instagram all proliferated as mobile apps – whereby individuals could download apps to their phones and share personal opinions across various platforms. It also allowed e-commerce to develop as a major digital force and increased the popularity of instant message services like WhatsApp, which were downloaded to millions of phones. Suddenly, apps offered users the opportunity to buy printer cartridges, order take-out, talk to friends and fabricate digital selves from a singular mobile device.

That brings us to the here and now, but what's next?

Technologists believe the fourth generation internet will move us past web browsers and apps and even beyond personal computers and mobile devices, to a cognitive network through which the physical and virtual worlds collide. This internet will be formed on the foundation of augmented reality technologies that are already emerging. Interaction will most likely happen through smart glasses similar to those already on the market today. (This is technology foreseen by those digital natives we spoke to in 2015.) Users will stop visiting web browsers and logging on to apps. Instead, the internet will be projected out onto the world in front of them.

We realised that the dawn of a new internet offered the Telco industry a colossal opportunity. Whenever a new version of the internet has emerged, the evolution has forced major organisations into obsolescence and allowed new companies to emerge in their place. Take the transition from the World Wide Web to the app-based internet as an example. Nokia, once a telco behemoth, became irrelevant when the world began to use smartphones – it was not ready to compete with Apple's iPhone. Yahoo became irrelevant when its competitors, chiefly Google, began to monopolise the search engine market, leaving them out in the cold. Neither Nokia nor Yahoo foresaw the stark changes the app-based internet would bring about, and they did not react quickly enough to remain relevant in the new app-focused world. At the same time, Apple and Google gained new prominence. The iPhone began to sell all over the world. "Google" became a verb.

Why did this happen? Well, certain CEOs recognised that for each iteration of the internet to have global success, it required a small number of companies to create a standard for use. And that the companies that created the standard often became all-powerful.

In the case of the World Wide Web, that meant creating a web browser and web page system that everybody in the world could access and understand. It didn't matter if you were in a hotel in Switzerland or at home in Seattle – the experience was the same. If your organisation facilitated that standard, your relevancy prospered. But if you didn't, you were toast.

In the case of the app-based internet, that standard was realised as an operating system – Apple's iOS or Google's Android – to which almost everybody in the world subscribed. The companies that created those standards, Apple and Google, became supremely prominent. The companies that were late to realise that standardisation was key – Nokia and Yahoo – became irrelevant. Think about it. Can you name a third mobile device operating system?

We realised that it wouldn't be too huge a leap to believe that the same will happen with the next generation internet, that a small number of key players will create an AR standard that everybody in the world will adopt. As the AR internet seeps into our everyday lives, some of today's main players will realise that they have missed the opportunity to contribute, and then a more damming realisation will appear: that they might not make it as a business.

In their place, new players will appear, or old companies that had the foresight to prepare, perhaps using a ten-year study that asked children to share their views of the future.

Deutsche Telekom, we realised, has an opportunity to define the standard for the AR internet, and to create a universal language that objects affected by this new technology could use. The opportunity would revitalise Telekom's position, not just in the telco industry, but in the international technology sector. We thought: why wait for Microsoft, Amazon, Google or Facebook to create the new Augmented Reality internet standard, when we can just do it ourselves?

*

For the past two years, we've been attempting to do just that. But it is a complicated process. On the one hand, we need to create a standard for the next generation internet. But on the other hand, we also need to convince numerous organisations to subscribe to the standard we create. Whatever we create will not be successful without uptake. What's the point in creating a language without an ecosystem of subscribers willing to commit to speaking it? Convincing other organisations to come on board is perhaps the most important element of this project. Without subscribers, the project fails.

Let's start with the first element, though: how to create the Augmented Reality internet standard.

First, we must split the standard into two halves. On one side, we have infrastructure: a new operating system that allows people to access the AR internet. This is the equivalent of the web browser.

On the other side, we have the language required to allow every connected object to speak to each other. In the fourth generation internet, augmented reality representations of all objects, even humans, will exist. Let's assume you drive a German car, a BMW, and you're on a driving holiday in France. One night, you approach a traffic light, the digital twin of which immediately connects with the digital twin of your car. The roads are empty, so the digital twin of the traffic light, knowing you are coming, turns to green, allowing you safe passage and preventing the need for you to slow down. And on you go with your journey. Digital twins already exist. General Electric has created digital twins of parts of aircrafts, leading to productivity savings of $125 million in 2016. Siemens has used digital twins to connect R&D, planning and production, subsequently reducing time to market for its products from a whopping 30 months to 16 months. And Black & Decker has created digital twins of its assembly lines, vastly improving operations.

But these digital twins speak different languages. The digital twin of a GE aeroplane part cannot, as of this moment, talk to the assembly line at Black & Decker. There is no digital twin ecosystem, no standard via which every digital twin can interact.

Our aim at Telekom was to create an alliance that defines the standard language that allows digital twins to talk, and a standard infrastructure by which objects can connect and that also allows the owners of those objects to control the data that gets shared.

Over the past two years, we have begun to develop both a distributed AR internet infrastructure and a universal language for the sharing of digital twin data. We envisioned it as an open ecosystem, a concept opposed to the idea that an ecosystem might be created by one dominant player. Subscribers to the system will have non-discriminatory access. It will be industry-agnostic. It will attract developers around a common platform. And it will be transparent. Crucially, data sovereignty will remain with the data's owners.

Last year, we began to share this concept with other organisations – potential subscribers – and sought to bring on partners in the creation of an infrastructure. We began to talk with manufacturing companies as well as software partners. Soon the German government offered its support. And then the European government did too. The approval is pertinent. In the past decade, large European organisations have suffered from the continent's fragmentation. That has allowed American software companies, particularly those from the Bay Area, to monopolise standards. Here's one example of how. The US has 35 weapons systems – guns, tanks, aircraft carriers – that work neatly together. In Europe that number increases to 178, and there is little interoperability between systems. The same is true among organisations.

That's why encouraging European organisations to embrace the Telekom created standard is so important. The more organisations embrace it – while simultaneously giving up on their singular ambition for ecosystem control – the greater the chance the ecosystem has of success. This is a huge challenge. Convincing organisations around Europe to give up their own attempts at developing a single standard and instead, to embrace ours, will be difficult. But here are the benefits of that system working: economic growth, societal improvement, the stimulation of innovation, transparent data ownership and control and a greater quality of life. It will grease the wheels of the sharing economy and help realise new business models. It will allow developers and entrepreneurs to use digital twin data to inform new ideas. It will give us choice. It will improve our health.

What will this all mean for the future of Deutsche Telekom? First, it will bemuse an infrastructure provider. It will provide the connectivity necessary to allow digital twins to interact. It will create secure digital passports for all objects in the ecosystem. And it will provide the low latency necessary to facilitate the AR internet. (The Augmented Reality internet will require high computing power to exist, which Telekom can provide.)

Most of all, this presents an opportunity for Telekom to become even more relevant, to create an internet standard that is used for the next two decades, to reach the level of Google, Amazon and Facebook. Isn't that exciting?

ESSAY:
PERSONAL INTELLIGENCE
AND DIGITAL TWINS

T he growing unease around the rapid rise of Artificial Intelligence has an
instrumental driver: we believe we are being somehow conned. In the
minds of many – not least a great number of tech luminaries, including
Elon Musk, Bill Gates and the late Stephen Hawking – AI is a dangerous technology that allows us to be secretly scanned, evaluated and surveilled. If it is
not quickly regulated, the thinking goes, what's to stop it destroying humans
and taking over the world?

Dystopian predictions like these have been central to countless sci-fi
fantasies, and they are unduly pessimistic. But it is true that significant effort
is required to reconcile people with machines, to help people understand
how disruptive technology might benefit us individually and as a society – to
demonstrate that AI, if treated correctly, does not have to become a threat.

This can only succeed if we change our priorities. It is important to ultimately achieve transparency about the impact of these technologies on society.
The focus cannot be on turning people into passive objects of rapid economic
and social change through digitisation. Instead, we have to empower them to
become self-determined agents of meaningful application and smart use of
these wonderful technologies.

If Artificial Intelligence can be transformed into Personal Intelligence with
the help of customer experience, Personal Intelligence will turn the individual

from an object of technology into a subject of free will and make them the sovereign master of their very personal universe.

This could have a big impact for democracy, society and economy.

Part of that effort involves creating Personal Intelligence, which would allow us to use 21st-century technology as a tool, while retaining autonomy over our lives. PI provides individuals with a digital twin over which they have complete control. It will store information about our relationships, attitudes, sexual preferences, political views, professional skills and a vast spread of interests, wishes and desires. It will use a personal interface to connect with other digital twins, as well as data repositories, and will only share the information we allow it to. At the same time, it will create value for us by acting on the information it compiles.

The digital twin is not a new idea. The concept of pairing physical entities with digital equivalents has roots in the early days of space travel, when NASA built scale models of spacecraft that, once launched, were far beyond their reach. The models were used to help monitor equipment onboard remotely. NASA's digital twins allow spacecraft to be monitored remotely during missions. Automobile manufacturers use digital twins to provide after-sale service. Airlines use digital twins of parts like wings to better identify potential flaws. Countless organisations use digital twins to bring products to market faster and at a lower cost, safe in the knowledge that they can try out an endless number of design iterations virtually, back in their factories, and make updates to real products later.

But what if a CEO could use Personal Intelligence to create digital twins of individuals? Not a corporate algorithm but a personal one. A program that works on the individual's behalf. That acts as their personal spokesperson, lawyer, consultant, teacher, lobbyist, headhunter and therapist. A personal buddy an individual can really trust.

Algorithms have already had a monumental impact on society. Consider the position of Google, Apple, Facebook and Amazon, multinational corporations that use Artificial Intelligence to market huge amounts of data for advertising, generating tremendous profits. They intelligently evaluate and exploit users and their information, transforming consumers into sellable assets. But the tide is turning. Consumers around the world are beginning to regain control of their data. They want to oversee the decisions they make, the organisations they interact with, of what they buy and who they buy it from. They suspect that with the rise of powerful computer algorithms, they have become marketable binary data, which is in some cases not far from the truth – and not the sovereign citizens they once were. In short, they want their power back.

Data security – the personal information we store online and who has access to it – has become an enormous point of political discussion. Legislation on data protection is being introduced around the world. Here, too, PI could provide solutions. Consider the value a CEO can create by creating a PI system that allows us to store our data privately. Wouldn't that prevent big tech from mining us for our data? Wouldn't it give us back control? Wouldn't it breed huge trust in and loyalty towards the organisation that introduces PI to the world?

Let's take another example: the jobs market. Some of society's most widely held fears are centred around employment. As AI becomes increasingly powerful, will we all lose our jobs? Will human beings face mass unemployment? Will robots become doctors, lawyers, rubbish men and baristas? This question is one we have asked before. During the first industrial revolution, as labour became automated and newfangled machines threatened to steal our jobs, society began to question its place in the world. If we didn't have a job, how would we make the money necessary to live? We needn't have worried.

Automation led to gains in productivity, which facilitated economic growth, which increased the number of jobs. Those jobs weren't the jobs we were doing before. Rather than weaving cloth ourselves, we operated machinery that weaved it for us. Machines took over mundane tasks, leaving us to concentrate on the sophisticated processes machinery couldn't handle.

Artificial Intelligence has stirred similar questions, namely: will an algorithm replace me? But do we really need to worry? Not if organisations shape PI into a tool that extends our abilities rather than replacing them. What if PI teaches us all how to develop our skills and interests? What if, by understanding who we are and what makes us tick, it feeds our curiosity with information it has compiled from the internet, our peers and data repositories? And what if the information PI has provided allows us to meet the requirements for our dream job?

Consider the value the CEO of a large organisation could create by developing Personal Intelligence. They might produce technology that serves as the foundation of a new, consistently successful global prosperity paradigm. Radical technology that is able to think from a human perspective. A program that not only benefits business in the products it creates and the sales figures it achieves, but that also acts responsibly, on behalf of society. This is the beginning of a shift that places human needs at the centre of enterprise. Tomorrow, the world's leading organisations will have moved beyond manufacturing products and services and instead will use Personal Intelligence to create deeper, more meaningful relationships with its consumers. Businesses and its customers will create partnerships. And everybody will benefit.

HOW TO APPLY PERSONAL INTELLIGENCE

In the digital twin internet, the next generation internet, physical and virtual reality will blur. The digital twin as a digital representation of objects and persons appears. As shown above – this is the successor of the current applications-based internet. In the history of the internet in transition phases like these new businesses have been born. Applying PI in this new internet creates extended opportunities for individuals and companies. As a business executive, product manager or design leader, understanding and leveraging the new opportunities around the augmented reality internet, the digital twin and PI will be key.

Taking on a radically human perspective, managers should not only think in products and sales figures, but start to develop a greater sense of social responsibility. Companies and brands are now moving closer to the center of people's lives. Traditional boundaries between businesses and personal life are dissolving.

Actions:

1. **Envision the position of your company in the next generation internet environment of the PI-based digital twin.**

2. **Identify the value drivers for your organisation in this potential future and how to leverage PI.**

3. **Ideate scenarios and options. Rapidly prototype, test and iterate future experiences, products and services. Putting humans first.**

4. **Get a systemic overview of the position of your company in its industry/environment. Graphically represent the interactions and interdependencies to other businesses, industries and their respective offerings as these things fuel PI.**

5. **Identify necessary partners and stakeholders horizontally and vertically for success.**

6. **Derive the necessary global software interfaces (APIs). Specifically understand the importance of a systemwide view. Define software interfaces and system APIs.**

7. **Define a product and service portfolio to reposition your business model from AI to PI (if you are already using AI).**

CHAPTER 7

TRANSFORMING FOR INNOVATION

Three types of organisations applying PI will appear:

1. Seeing digital projects as a pure marketing tool for brand value.
2. Focus on efficiency – progressing existing business linearly. Seen often still as digital heroes.
3. Disruptors or creators of new business. For these exceptionally few, Personal Intelligence will transform the entire organisation. Coping with ambiguity will be key for strategy and culture.

The conflict between rigid hierarchical structures in the core and the need for fluidity leads to conflicts. Leaders need to combine two models: First, the blue geese model – the current organisation – one geese leads and the other follows. Second, the green flock model – a flock of birds dynamically adjusts to the environment. And can easily outmanoeuvre an attacking hawk. In the flock model each bird is directed by the purpose of the overall organisation or its sub-unit. Within the second model the organisation will lose its weight and gain lightness. Its processes and structures as well as its products, services, experience – literally everything, even solid objects like stones – will become intelligent and will feel alive. Which will add to the fluidity of life.

Organisations combining both models – called ambidextrous blue-green organisations by Stanford professor Burgelman – will thrive.

Top actions for the CEO:

1. **Assess which options the digital transformation has in store for your specific organisation. From innovation marketing to disruptive transformation.**

2. **Decide how much risk you want to take. And choose carefully between the options. From innovation marketing to disruptive transformation. Resulting in a clear direction for innovation.**

3. **Devise your strategy. Include all parts of the organisation: blue and green. And their specific needs of hierarchical structure and fluid environment. Identify culture change agents. Move them to positions of power.**

CASE STUDY: CREATING A CREATIVE ENVIRONMENT

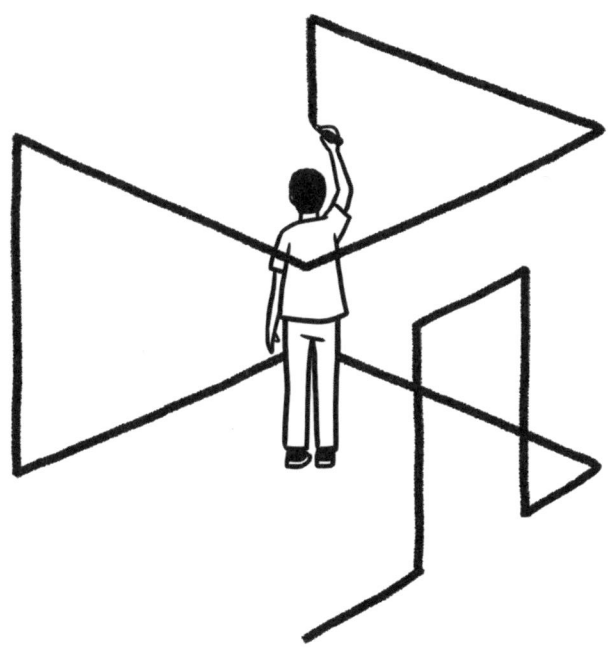

To adapt to the challenges of transformation and create a truly innovation-centred culture, Deutsche Telekom realigned its way of working around innovation. A major activity was to introduce a Portfolio and Innovation Board (PIB), responsible for managing the roadmap and coordinating innovation projects across the Deutsche Telekom group. The board met monthly and was co-chaired by the CPIO (Chief Product and Innovation Officer) and the Senior Vice President for strategy. Participants came from top management, representing different innovation, marketing and technology functions.

Taking control of Deutsche Telekom's myriad and scattered digital innovation investments, the PIB created greater transparency around investments and results. Involving both P&L leaders (business leaders with Profit & Loss responsibility) and Deutsche Telekom's C-Level (executives reporting to the CEO), the PIB also engendered greater "buy-in" to the innovation portfolio from Deutsche Telekom executives, and a more collaborative style of working together across functions – particularly the innovation, marketing and technology functions. Innovation was no longer limited to a central unit: anyone with an idea was encouraged to submit proposals, working with a member of the PIB.

To incentivise P&L leaders to also collaborate, the CPIO set up the Corporate Innovation Fund (CIF) to fund strategic opportunities that emerged outside of the annual budget planning process. It created an opportunity for P&L leaders to start new projects and areas without direct impact to their P&L. Result were win-win-wins for the business, the individual business unit and the individual leader. This made it attractive for P&L members to collaborate – leading to emotional buy-in.

The financing process of the CIF was based on a venture capital approach. Projects were pitched to the PIB during one of its monthly meetings; approved projects received funding for a single project phase (e.g. about 2–3 months), comparable to a Series A round. At the end of a project's funded phase, the PIB assessed whether the project had achieved predefined milestones and whether it qualified to receive additional funding, a quasi-Series B round.

This process was designed to encourage project teams to propose and deliver realistic outputs, and to provide greater flexibility for the PIB, enabling the board to stop and reprioritise projects based on lessons learned and remaining funds. Previously, this type of assessment and prioritisation only occurred on an annual basis. Project funding was limited to one year, after which a project team was expected to find a business owner for its innovation project, folding the project into Deutsche Telekom's annual financial planning cycle.

As the growth of mobile phone revenue slowed, Deutsche Telekom had established a Digital Business Unit to create new business. However, the desired business impact did not surface. Many large corporations experience the same challenges when driving innovation. Research shows that core businesses tend to protect themselves against disruptive innovation.

As a restart, a central innovation unit called Group Innovation was created. Group Innovation was positioned to provide "innovation you can trust". Trust to deliver value for the end customer, value for the business unit leader and value for team members as their time was used wisely. In summary – creating emotional buy-in by delivering a win-win-win approach differed starkly from the advice in popular business literature to create standalone innovation units shielded from the rest of the organisation. This had been the precise problem with the Digital Business Unit.

Emotional engagement is key for success. Emotional engagement, however, cannot be decreed from the top down. Instead, we empowered a group of informal thought leaders to search for new approaches to innovation in a flock approach. They conducted interviews with 20 project and topic leaders and studied the latest academic research from around the world. Moreover, the task force ventured out and spoke to other corporate bodies, venture capitalists and startups that included coworking spaces.

In response to these lessons learnt from the limitations of previous approaches, Product Innovation organised its transformation efforts around three principles:

1. Focus: Reduce the areas of innovation that Product Innovation will pursue in the future.
2. Flexible Funding: Take a more flexible approach to fund projects.
3. Collaboration and Partnerships: Ensure all innovations are owned and executed jointly by the central innovation unit and business segments and that the innovations draw on external experts and partners.

To communicate focus on a few key innovation areas, the team created an internal brand: "4+1": four major innovation projects with short-term impact would be executed, plus one medium to long-term project.

In practical terms, innovation areas were reduced from 64 down to these 5. As a result, several innovation projects were sold or discontinued.

Such a reduction would be helpful for other organisations as well: let's assume there are 40 stakeholders for a given topic. Their approval and support are required for the innovation project to succeed. Now assume a 90% probability of approval by each of them – the resulting likelihood for success is (0.9 * 0.9 * 0.9. 40 times) % = 90%^40 =1.3%. Resulting in the need to focus due to stakeholder management. We derived the "von Reventlow law of innovation in large corporates": the more stakeholders exist, the less likely innovation becomes. Or inversely formulated – with a growing number of stakeholders, more focus and leadership are required. Plus, the necessary effort to create emotional buy-in for everybody through win-win-wins grows substantially.

The ambidextrous approach to leadership and organisation was applied. The incremental four projects were called blue – incremental changes to Deutsche Telekom's core business, generating added value that could be measured in terms of euros and customers. Green were more disruptive innovations with a longer-term focus. Although their benefits were significantly more uncertain, the green innovations had the potential to mark a turning point, radically change markets and disrupt one of Deutsche Telekom's current business models. A green innovation was most likely to come from R&D efforts outside the company.

To speed up decision-making, the small Investment Committee was established, comprising the CPIO and the Finance Senior Vice President of Technology and Innovation. This approach provided project teams the room to breathe; it's lean and simpler as group-typical steering committees acting like an internal venture capitalist.

The Investment Committee was instrumental in ensuring that projects could "fail fast and cheap", i.e. be stopped in time to mitigate negative impact. Previously, innovation funding followed a more traditional waterfall process.

The Innovation review had shown that innovation projects were being hindered by conflicting interests. To circumvent this, we introduced a new federated approach: all innovations would be led clearly by both a senior-level manager from Product Innovation – responsible for evangelising push innovations – and by two senior-level managers from a business segment – responsible for market aspects. Shared ownership again created buy-in and the experience of win-wins for everybody involved.

The results were visible and corroborated by an independent MIT case study: top management was spending significantly more time on innovation, actively participating in projects and in committees, assessing project viability, and orchestrating funding and synergies across projects. These improvements were apparent both in management's participation and in comments members of management posted on internal blogs remarking how much they preferred and enjoyed the new approach.

EMPOWERING DESIGN AS CULTURAL CHANGE AGENT

The design department played a decisive role as cultural change agent by creating a transformative environment. A cultural change agent is defined as driver, role model, thought leader and facilitator for new ways of thinking and doing.

Leading businesses have realised that innovation is not just about technology – it's about what people need. Therefore our approach to position the design department as mediator between human needs and technological possibilities. We combined design thinking with technological advancement and created tangible visions of the future. We embraced design as a core activity in processes and decision making.

To be truly effective, design needs to come from a design-minded organisation. That's why we worked hard to develop the strategic design skills of our colleagues and drove the digital transformation of the company. To equip the organisation with the right design tools, we created a framework of methods, tools and approaches. And we use them to support design thinking throughout Deutsche Telekom. Because designing a seamless customer and brand experience is a joint effort – it cannot be achieved by designers alone.

Telekom Design maintained close ties with international designers and design institutions. In such conversations we wanted to learn from the best, but also reflect on our own experiences and recognise trends and discourses that are inspiring and relevant for the digital transformation of our own organisation. This is an existential matter of survival for any market-oriented company in the world, especially in view of the major challenges posed to the economy and society by digital transformation and Artificial Intelligence.

An open dialogue with experts from all kinds of disciplines is crucial to discover the cultural developments in societies and to relate to the worlds of diverse consumer groups. Inspiration from the outside is the most important precondition for internal changes.

Receiving recognition for our transformation performance from the international design and digital communities is especially motivating for internal change agents from all departments of the organisation. Furthermore, it improves the company's overall reputation and, most importantly, it increases Deutsche Telekom's appeal as an employer for the best design talent.

Telekom Design's reputation across the globe is driven by the outstanding number of international awards it receives. Regularly participating in prestigious international and national design competitions puts a spotlight on the innovative, customer-oriented and aesthetically sophisticated design of Telekom products and services. Telekom Design is state of the art and lives up to its claim "We Design Simplicity".

The consistent improvement of its customer experience garners high recognition worldwide. In the past four years the design department at Deutsche Telekom won about 200 international awards and ranked 26 in the creative ranking of the Bundesverband Digitale Wirtschaft (BVDW) e.V. This made Telekom Design one of Germany's best agencies for digital creation and the only in-house design team of a corporation to be part of the German top 30 digital agencies.

The team committed itself fully to the cultural changes in economy and society. It entered a systematic process of professional development and embraced the question of how the digital transformation will change the way people live, work and communicate in the future. As a team, they cultivated a collective design mindset.

The internal partners experienced the outcomes – it was both intellectually as well as emotionally satisfying. And furthermore, we democratised the process of design as a cultural mindset. Meaning creativity, customer centricity and constant iteration for the best tangible results as a core trait in the DNA of the organisation.

A design academy was established in connection with HR and other initiatives around design thinking over the years. It delivered more than 8,000 design thinking trainings in the first two years. Turning theoretical knowledge into practical action, the Design Academy supported colleagues with their day-to-day work life and trained their employees and project teams on how to successfully apply design thinking methods. The academy joined the internal clients searching for innovative ideas, whether it's for a short-term project analysis, or campus formats over a period of several weeks. To understand the customers, they included them in all our development phases. This inclusion took place in formats like Customer Safaris and in-depth interviews, focus groups and co-creation workshops as well as Customer Labs and international feedback surveys.

The Customer Labs was the place where our customers participate in every phase of the product and service development. From the very beginning of the process, to the launch, and beyond. By involving customers, we gained valuable feedback that enabled us to refine our prototypes and eventually turn them into relevant propositions for our customers.

The "Telekom Customer Personas" – real customers representing the archetypes of our segments – were present with every step of the process. In this way, research was enabled along the design process. All necessary research methods for every phase of the design process were taught and mediated at the Design Academy.

To establish a culture of innovation, the Telekom Design Academy provided the opportunity to interact with each other. The design thinkers of the group took part in events and meet-ups, shared ideas via the intranet platform, and learned about new methods and impulses from other companies. This strengthened our method within the group DNA.

As part of the work we tailored a holistic framework to enable every-body at Deutsche Telekom to work in a customer-centric way and with the latest methods and tools. We published the book "Design Thinking Doing" – a printed toolbox, a selection of the best design thinking methods with processes, detailed personas and practical tools – all custom tailored for the Deutsche Telekom. We mapped design thinking methods to our standard development processes to show which methods create the biggest impact. We made it as flexible as our work culture. Instead of printing an unchangeable book, we made it a slipcase with an integrated binder, so it can be continually extended and updated. We added an online platform and processes to ease daily use. As beautiful as our slipcase looks and feels, we still wanted to make this standard set of methods and tools available to anyone at any time.

A large variety of formats of engagements with customers was invented. Resulting in improved customer experiences, processes and cultural shift. In summary, this was highly valued internally and externally. Constant positive re-enforcement due to the successes established a new alternative behaviour path for the organisation. Culture got transformed.

ESSAY: CHANGE AGENTS FOR DIGITAL TRANSFORMATION

I nnovation is critical for the survival of any organisation. That is especially true today, as technology expands exponentially and algorithms become both more intelligent and more prevalent. To be unique, businesses must strive for innovation. But it won't be easy. It is not in the nature of large organisations to constantly reinvent. Actually, the opposite is true. Reinvention interrupts highly complex systems that have been developed over years, and can cause significant drops in productivity and quality. That's risky for any business.

But it doesn't have to be. In order to prosper, organisations must place faith in the people they employ. People are the real assets of any company. It is their ability, creativity and collective engagement that make a company what it is. When employees commit their hearts and souls to the organisations for which they work, those organisations thrive.

Employees of innovation-driven corporations need a high degree of liberty, and they need to work in a supportive creative environment. Designers and innovation managers as cultural change agents enable organisations to inspire all employees towards innovation, to give them the scope and the tools to dare to think creatively and disruptively – to be open to change and to be willing to take risks, even if it means ending in potential failure.

To develop a creative environment, the first step is to remove bureaucratic obstacles and create a climate that promotes innovation. Without employees who feel supported and free to develop ideas, innovation will not thrive. This is why employees of innovation-driven corporations need a high degree of liberty and in particular free access to common knowledge resources like they exist in communication networks, databases and cultural groups.

Plus, they need:

1. Far-reaching autonomy regarding how they work (and what they work on).
2. Possibilities for collaboration and co-creation across disciplines and functions.
3. A culture of trust.

It is applicable for large-scale corporations that ideas are literally in the air and born by the thousand, yet they die like fireflies if not conducted through a systematic innovation process and supported from idea to marketable product with appropriate talent at every relevant point. A process of innovation in a large corporation is extremely multifaceted because it is integrated in complex correlations of production and international co-operation with customers and suppliers. An open yet efficient and agile innovation process therefore has to continually be fed back, rethought, corrected and revised.

In our experience, it is essential to revise the innovation process from the ground up, starting with the corporation's core capacities: only in this way will existing experience and knowledge be mobilised, and only in this way can practical knowledge be radically questioned.

Customer needs are even more important within the innovation process. Customers have problems and are searching for solutions. Problems are nothing negative here. Anyone wanting to create something new has problems. A company does not always have to love its customers, but it has to love their problems. These problems can only be solved by engaging in a close iterative partnership with the customer. As a result, products will actually be successful in the market.

If innovations drive market success, what drives innovation in the company? Design! Unlike any classical consultancy, design has the empirical basis and the creative fantasy to predict the future. Because design recognises customer needs. Because it is able to anticipate scenarios for applications. And because it can translate such scenarios, in an agile process constantly adjusted to customer needs, into prototypes. On the outside, design builds meaningful connections to customers. On the inside, it is the key to strategically control innovation processes within complex structures. Not in a conventional sense of design as giving products a distinctive shape and humanising interfaces between technology and man, but as a cultural change agent.

Design as a cultural change agent enables organisations to inspire all employees for innovation. It provides the employees with the scope and the tools to think creatively, to be disruptive and open for change, to take risks and to have the courage to fail. Often called design thinking, design as cultural change agent has moved from the level of products to that of organisational forms and processes. It has taken on a leading role in the business strategies of innovation-driven corporations. When design becomes a relevant part of business strategy, it carries leadership function.

Design leadership implies that a CEO and their management board have to systematically deal with the transformation of technology and society. From these insights they develop possibilities that are innovative as well as applicable and then discover opportunities for growth. With creative intelligence, they identify trends and talents and combine them with the momentum of their organisational power.

Actually, these are the traditional virtues of the entrepreneur, whose role it is, according to Schumpeter, to introduce innovations. Schumpeter says: The entrepreneur is an innovator who is driven by the joy of being creative. Therefore, a CEO wanting to sustainably increase his business value has to shape new values with the creative potential of his company. Achieving this, he will become the leader who applies his convincing visionary strength to inspire the employees to follow him and conquer new profitable market opportunities.

HOW TO TRANSFORM FOR PERSONAL INTELLIGENCE BASED INNOVATION

Personal Intelligence based innovation starts with seeing the human as the centre of all considerations, combining AI and UX – creating new tools for customers, employees, corporations and general humankind. Starting with the human in mind, we approach this transformation in a two-pronged strategy – creating both financial as well as cultural win-wins in each activity:

1. From today looking forward to the future. Focus here is to create one or two quick wins as proof points. Creating momentum for the transformation to PI.
2. Retro casting from the future to today. Looking backwards from a potential target picture and deriving what would need to be done today. And implementing again one or two examples as proof points for the value created through the longer-term journey.

And a frontrunner of cultural transformation for the ambidextrous organisation must be internally identified. If the organisation is willing to transform sustainably, the design function can play this central role as the catalytic engine for the PI cultural journey and create the narrative of the future. This translates into embracing design as a core activity in the innovation processes and decision making, to create value for the customer, the stakeholders and the company.

To be truly effective, design needs to come from a design-minded organisation. That's why it's necessary to develop the skills of all employees to successfully drive the digital transformation of the company. A central education unit combined with effective formats for customer engagement like personas, service safaris and customer labs can help to increase the customer centricity of the whole organisation.

Finally: staging and celebrating wins for the customers, the businesses and every stakeholder creates positive re-enforcement, driving an accelerated journey.

Actions:

1. **Assess which options the digital transformation has in store for your specific organisation. From innovation marketing to disruptive transformation.**

2. **Decide how much risk you want to take. And choose carefully between the options. From innovation marketing to disruptive transformation. Resulting in a clear direction for innovation.**

3. **Devise your strategy. Include all parts of the organisation: blue and green. And their specific needs of hierarchical structure and fluid environment. Identify culture change agents. Move them to positions of power.**

4. **As kickoff for the transition to an innovative organisation, create an emotionally engaging experience for the top 50.**

5. **Provide the necessary protective space for creatives and change agents and ensure they can rely on strong backing from top management.**

6. **Create spaces for your creatives and designers where they can work and express themselves freely. Open up the organisation and strengthen its interdisciplinary capabilities, so it can productively partake in the discourses around design and digitalisation.**

7. **Establish a design mindset. Make use of designers' special talent to glimpse social trends and cultural changes and to empathise with consumer groups and their various worlds.**

8. **Take each and every person with you on the journey. Have them self-select their informal leaders. And have the informal leaders drive the transformation process.**